KEEPING CHICKENS

Practical Advice for Beginners

KEEPING CHICKENS

Practical Advice for Beginners

9th, revised edition

50 colour photographs
52 drawings

Beate Peitz

Leopold Peitz

Translated by David Adams

First published 1985
This edition published by 5m publishing 2016

Original title: Hühner halten

David Adams asserts his right to be known as the translator of this work.

Published by
5M Publishing Ltd,
Benchmark House,
8 Smithy Wood Drive,
Sheffield, S35 1QN, UK
Tel: +44 (0) 1234 81 81 80
www.5mpublishing.com

A Catalogue record for this book is available from the British Library

ISBN 9781789184129

Book layout by
Keystroke, Neville Lodge, Tettenhall, Wolverhampton

Photos by iStockphoto/anzeletti, Rainer von Brandis, Tony Campbell, Pamela Cowart Richman, Tatiana Fuentes, iStockphoto/johnnyscriv, Dr. Heinz Linke, Linda Steward, Studio Annika, Silke Klewitz Seemann, Regina Kuhn, mauritius images, Beate and Leopold Peitz, Reinhard Tierfoto, Friedhelm Volk

Illustrations by Rainer Benz and revised by Helmuth Fluhbacher

Contents

Foreword

Keeping chickens is often the only possibility, in our densely populated environments, of involving oneself closely with livestock and bringing a little bit of real country life – surely a dream of many in our time – into one's own garden. Although the keeping of chickens has mainly developed in two very different directions in recent decades, firstly, in commercial poultry operations with huge numbers of high-yield animals and, secondly, in specialist breed keeping with small numbers in a wonderful variety of colours and shapes, it is clear today that the number of small mixed-breed chicken populations is increasing and also that some old breeds which were once kept in the country are being rediscovered and protected against complete disappearance.

For the latter group of keepers, alongside the idea of growing their own produce, there is also joy in the creature itself and an immediate sense of closeness to the cycle of life and death. Some readers, because of their circumstances, will never be able to enjoy this enriching experience, while others may lack the courage to take on the challenge. This book is particularly dedicated to people in both of these categories. To the one, in the hope of extending their knowledge and understanding of the living environment and to the other to give them the specialist knowledge they need and encourage them to dare to try. We will try to do this by showing, starting with the cultural history of the domestic chicken and its physical and social characteristics, what we mean by humane husbandry which, we believe, requires more knowledge than is commonly put into practice.

And so, we hope the would-be chicken keeper will feel able, given the knowledge gathered in this book, to judge with confidence under what conditions – given a firm resolve – he could turn his idea into reality. Whether it is the question of acceptability to the local municipal authorities or to next-door neighbours of keeping a modest chicken flock; or the problem of constructing a suitable hut, the design of a run, correct care and proper feeding; whether issues relating to artificial or natural breeding and rearing or ideas for using chicken products; all these should find answers here.

The authors have no wish to lay down dogmatic rules about chicken keeping. Rather, their intention is to provide inspiration and to offer guidelines, while leaving room for the reader's own individual, creative solutions.

Beate and Leopold Peitz

CHAPTER 1

The Domestication and Spread of the Chicken

It is sometimes not easy to decide just where a story should start, but we begin our cultural history of the domestic chicken as recently as the second millennium BC. A proper understanding and appreciation of the chicken and its products as a food source – so widely undervalued today – calls for some knowledge of its cultural and historical origins and the transformations it has undergone throughout the development of human civilisation. Equipped with this knowledge, we will soon begin to see our little flock of birds through different eyes.

A Brief Cultural History

Speaking of chickens today, the average person will probably think first of hens laying eggs and then perhaps of a proudly strutting, crowing cock, an image which has actually had far greater cultural significance than that of the female of the species. Among many ancient peoples, the cock was regarded as a sacred being.

It is believed that the ancient Persians held the cock in high esteem because, in their view, he drove away demons and sorcerers with his crowing and, for that reason, he was made the protector of the household and its livestock. Later, this living mascot was brought, in the course of the Persian wars, to Asia Minor and was subsequently adopted by the Greeks who honoured him with the name *alektor* (meaning "protector").

Later on, the cock came to be used as a sacrificial animal among the Greeks, offered up mainly to Asclepius, the god of healing, by way of thanks when a person recovered from an illness.

Among the Romans, chickens acquired another role, which was first ascribed to them by the Greeks in southern Italy.

It was believed that the chicken possessed the sacred ability to predict the future. Particularly when a military campaign was being planned or a decisive battle was about to start, an *auspicium* or divination was performed. For this, the

Page 8: It is not essential to decide on just one breed. A mixed group is also very attractive and can bring the same benefits.

specially appointed chicken keeper, or *pullarius*, scattered feed in front of the sacred chickens *(pulli)*. If the animals ate hungrily, this indicated a favourable outcome to the battle, but if they ate unenthusiastically, it could be concluded that the enterprise would end badly.

And it was not only for success or failure in battle, but also to help decide on other plans and even national policy that the chicken *auspicium* was made use of. As Pliny the Elder wrote, in his *Natural History*:

> *The cock is worthy of the honour that is bestowed on him by even the Roman consuls. His more or less eager feeding provides the most important information regarding the fortunes or misfortunes awaiting the Roman state. Daily he rules our rulers or closes and opens their own houses to them. He orders the Roman consuls to proceed or to stand still, orders or forbids battles; he has heralded all the battles fought on earth, he governs the governments of the world and, offered as a sacrifice, is an excellent means for winning the favour of the gods.*

But it was as a sacrificial animal that the cock gained ever greater importance among different peoples, particularly among the poorer social strata who were less able to afford to sacrifice larger animals. Then over time, resistance to using chickens for more worldly purposes such as egg production or even for eating gradually subsided. And as a result of this, we can find a great deal of guidance on the keeping of chickens in Roman literature. For example, Varro writes:

> *If it is desired to keep 200 domestic chickens at a country house, they should be given a separate hut, the space in front of it, on which there should be sand for bathing, should be fenced in and they should have their own keeper. If the eggs are to be kept for the kitchen, they are rubbed with powdered salt or place for three hours in salted water, dried and then covered with bran or chaff. If domestic hens are to be fattened, they are enclosed in a dark, lukewarm place and force-fed with boiled barley. Each time they are fed, the head is also cleaned of lice if necessary. They should be fat in 25 days. Some also fatten in 20 days and produce a tender flesh by being fed with wheat bread softened in a mixture of water and wine.*

The first-century Roman writer Columella, in his book on agriculture, provides detailed instruction on chicken keeping which, in its essential features, differs little from the modern practices of hobbyists and smallholders. He recommends, among other things, that the hut should be placed close to the kitchen or bake house so that the smoke can be let in to protect the fowl against disease and other dangers.

It is also interesting that even at that time, differences in the laying performance and meat yield of different breeds were noted. The Romans, who drew their chicken breeds from the Greeks, particularly favoured the birds from Delos, Rhodes and Melos due to their size and their high level of egg production.

Good to know

The reader might well notice that the ancient keepers of chickens obviously observed their animals very closely to investigate their needs. They clearly took the view that the wellbeing of the animals was only achieved through the most humane possible keeping and that this resulted in a greater economic viability.

The chicken is believed to have reached central and northern Europe independently of Greek and Roman cultural influences. Evidence for this comes from the Roman invasion of the Celtic lands on the south coast of England, about which Caesar wrote that they already had domestic chickens. The chicken was clearly sacred to the Celts and also to the Germanic peoples. It was seen as sinful to eat the chicken itself and it was, at best, permitted to consume its eggs.

It was only in the Middle Ages that the chicken gained its full significance as an egg and meat provider. Raising chickens then became a very important cultural and economic activity. As in so many fields, it was the monks in the monasteries who had the greatest success, and passed on the animals they reared to serfs and free peasants who, in turn, paid their dues to the manorial lord with chickens and eggs. At the time, chickens also constituted the favourite food supply for military campaigns since they could be taken along by armies, alive in large wooden cages. And later on, a flock of chickens became an indispensable part of every farmstead, the birds feeding themselves from threshing waste, seeds of all types, worms, insects and kitchen waste, and could therefore be kept cheaply as a side-line.

Other cultures were also won over by the domesticated chicken, spreading out from its Indian origins where it was first kept some 4000 years ago. On its journey westwards via the Near East and Asia Minor, it also reached Egypt and North Africa, although earlier its eastward spread had taken it to China and from there to Japan and Mongolia. It was only some 500 years ago that it arrived on the American continent and later still in Australasia.

As to why humans originally domesticated the chicken, two important factors can be identified which possibly preceded its role and a food provider.

When chickens were first put in captivity, they laid no eggs and failed to reproduce, and it is therefore the view of some historians that the usual reasons for keeping them, which seemed as obvious in the Middle Ages as nowadays, did not apply. Rather, it is suspected that chickens – or more likely, cocks – first aroused interest because they enabled people to put on cock fights, famous in their place of origin although abhorrent to us. Another possible reason is the cock's natural habit of acting as an alarm clock, unfailingly heralding the approaching morning with his crowing.

We may never have absolute certainty about which of these two reasons was decisive for the domestication of the common chicken, but we can be sure that the cock made a very early appearance as a joust-fighter for the common man, as he still does today, and that in antiquity and the Middle Ages, he was valued as a timekeeper, a quality that is no longer appreciated by modern man. This function remained even into the 20th century in the Orient where camel caravans always carried a cock with them so that they could wake up early enough to break camp and travel during the cool hours of the new day.

It seems evident that the cock gained great mythological significance in ancient times because of these behaviours. And even today, his image is found on some old church spires. Nowadays, however, he serves at best as an ornament or a wind vane and no longer as *alektor*, the defender against evil spirits.

But in his place, his female counterpart has acquired a far greater economic importance as a provider of eggs and meat. Latter-day animal welfare issues have actually given him back a certain degree of symbolic fame.

Further info

Further evidence that the chicken found its way to the Germanic peoples and the Celts independently of the Romans lies in the Gothic word for cock: hana, and for hen: hôn.

These powerful examples of the Cochin type clearly feel comfortable in this idyllic setting between the pasture and the manure heap.

In terms of cultural history, the cock's day is over, but now his wife the hen reigns supreme.

Further info

The size of the red junglefowl is roughly that of our typical miniature breeds. The cock crows like our domestic rooster and, mated with the domestic hen, generates offspring with unreduced fertility.

Origin Theories

Red junglefowl, grey junglefowl, Sri Lankan junglefowl and green junglefowl are names that are not familiar to everyone interested in chickens, but these are assumed to be the ancestors of our modern-day domestic chicken breeds.

Darwin, who said that he had kept, bred and researched almost all the English breeds of his time, considered it certain that the domesticated chicken was descended from the red junglefowl *(Gallus bankiva)*, which is indigenous to India. His main argument was that only crossings between the red junglefowl and the domestic chicken were fertile. Today, we know that crossings with other wild chickens can also result in fertile offspring. However, the red junglefowl is considered to be the main wild ancestral form among four possible ancestral forms. Other forms that are considered to be ancestral are

- the grey junglefowl *(Gallus sonneratii)*,
- the Sri Lankan junglefowl or Ceylon junglefowl *(Gallus lafayettii)*, and
- the green junglefowl *(Gallus varius)*.

The red junglefowl is found in the forests below the southern slopes of the Himalaya mountains, in mainland Southeast Asia and the Sunda Islands and feeds on insects, larvae, buds, worms and all types of seeds. The hen lays 8 to 12 eggs in a depression in the ground which she lines with grass and foliage to make a nest.

The geographic ranges of the three other parent forms are Sri Lanka for the Ceylon junglefowl, the southern part of the Ghat mountains and plateau of the Indian peninsula for the grey junglefowl and Southeast China and the Malaysian islands for the green junglefowl.

CHAPTER 2

Chicken Breeds Suitable for Small-scale Keeping

A Little About Breeds

The roughly 150 breeds of chicken that are known today – as with the large numbers of breeds of other animal species – are testament to the ingenuity of humans. After all, apart from purely economic motivations, the joy of doing the work of creator surely plays a large part in the modern diversity of appearance of our domestic chicken. So as to categorise this variety and to place it into a manageable structure, it is necessary firstly to differentiate the types as follows:

- large breeds without special characteristics,
- large breeds with special characteristics, and
- true and artificial miniature types.

Within these three groups, a general distinction is drawn again between three different types, based on their external appearance:

1. The Bankiva (red junglefowl) or farmyard chicken type

This type is characterised by a square, somewhat cylindrical body shape with a flat back, having the neck and tail set at an angle to it; white ear patches; smooth, close plumage and medium-length featherless legs; the cock's tail is long and curved; the skin and eggs are white.

2. The Cochin type

The Cochin has a massive body with a short back and a broad rump cushion rising toward the tail; the tail is short, stocky and slightly curved; the plumage is well developed and thick to bushy; the ear patches are red; the body skin and the egg colour are yellow.

3. The Malay type

The body of this type resembles its product, the egg; the breast is carried high; the plumage is hard and close-lying; the back slopes down and is narrow at the saddle; the neck is very long and is held upright; the legs are very long and feather free, with strong, prominent spurs; the ear patches are red and the body skin and egg colour is yellow.

Intermediate Forms

In addition, numerous intermediate forms exist today, having more or less pronounced influences from these three different types. For our modern farmed poultry breeds, the two types described under 1. and 2. are probably of outstanding importance, whilst the Malay type embodies a special form as a fighting animal. These features, inherited unchanged for centuries, have their effect in the bloodthirsty cock fights that, here in the West, are generally regarded as unacceptable.

Red Junglefowl.

This magnificent male bird is unmistakably a representative of the farmyard chicken type.

If we ignore, for now, the Malay type, which is not particularly significant in the Western world, among the breeds of the two other types, we can identify a large number of shapes and colours; for example, within the breeds of the Bankiva type there are the single-combed and rose-combed farmyard chickens.

The former includes, for instance, the Italian breeds and the world-renowned Leghorns, whilst the rose-combed type includes, among others, the Hamburg (or Hamburgh) and the Rhinelander. In the field of commercial poultry rearing, they are included among the "light layers".

Pure Cochins play only a minor role among our farmed chickens; of significance, at best, are well-known breeds within the Cochin type, such as the German Langshan, the Brahma and the Orpington. These exceptionally massive animals with a large meat yield are commonly included in the category of "heavy breeds". We can also find a good deal of their bloodline in the "medium-heavy breeds" which now represent, both in numbers and economically, the most significant sector in farmed poultry. These include, for example, the well-known Rhode Island Red, the Plymouth Rock, the New Hampshire Red, the Wyandotte, the Sussex, the Blausperber and the Faverolles.

And not to be omitted at this point are the numerous forms of mutated varieties and bantam breeds. The former include crested chickens, hooded and bearded chickens, naked necks, curly feathered chickens, Silkies and others. Like many of the miniature breeds, they are the pride of hobbyists and specialist fowl breeders. The miniature breeds have come about through crossing of large breeds and true miniature varieties (natural miniatures); in addition, weakly growing types have also been selected for cross-breeding. Therefore, many breeds exist which have both a large fowl form and a miniature form.

Good to know

Among the miniature breeds, a distinction is made between original and artificial miniature types. The proportion of true miniature types is very small.

Suitable Breeds

For the reader who does not yet have experience with chickens, the above division into types and breeds might seem, at first, to be a little confusing. The decision about which breed from this rich genetic pool would be the most suitable should be made based on practical grounds. To make it, the reader should consider how he wishes to make use of the animals, what space is available and which breed best meets his personal taste from the viewpoints of temperament and outward appearance.

It is reasonable to assume that essentially all the breeds of chickens kept in our latitudes – including some practically forgotten country breeds – are suitable for keeping in housing with a run: they are sufficiently weather resistant and usually willing to find part of their feed for themselves outside. Against this background and given the almost 150 breeds, only a few types are examined below, by way of example.

Helpful tip

For an overview of currently recognised breeds, the local breed clubs and small animal breeding societies can offer relevant literature and information.

Selection According to Utility

Eggs

Anyone who places more value on abundant eggs than a hearty Sunday roast and wants a light, lively hen with a strong drive to move around, should choose a light laying breed. As mentioned before, these include the Rhinelander, which originates from western Germany and has existed as a breed for more than a century. It is exceptionally weather hardy. In the breed description, it is said to be a typical tough, stocky farmyard chicken type with a rose comb. A hen can be expected to lay 180 white-shelled eggs in the first year, 160 in the second and 130 in the third.

A classic farmyard chicken type, as illustrated in many children's books, is the partridge-coloured brown Leghorn with a single comb. It has a somewhat nobler appearance than the Rhinelander but is not so robust as the latter. It is more suited to the moderate climate of central Europe. Laying performance is roughly as good. Other known representatives of the light farmyard chicken type are:

- Altsteirer
- Brakel
- German Sperber
- Hamburgh
- Kraienköppe
- Lakenvelder
- Leghorn (a purely commercial breed)
- Minorca
- Ostfriesische Möwen
- Thüringer Barthuhn
- Westfälische Totleger.

Shared characteristics:

Light body build, white ear patches, yellow legs, low brooding drive, white eggs.

Meat

If what we want is tender white meat, we need to consider the heavy varieties of chicken. Most breeds in this category belong to the Cochin type and have a very impressive appearance. Final live weights of 5–5.5 kg are not uncommon for fully grown cocks. Hens are also no lightweights at up to 4.5 kg live weight. A roast this size provides a good meal for a family of four or five and is no comparison with our supermarket chickens, which are basically conceived for a one- or two-person household. It is not only the Cochins and cross-breeds resulting from them that have brought these giant birds into being. A mature Dorking cock, an English breed also with a typical farmyard chicken appearance, can weigh up to 4.5 kg. The heavy varieties naturally also lay eggs, but in much smaller numbers than the lighter breeds. Exceptionally, Dorking eggs are white in accordance with their origin (they are of the Bankiva type), whilst in the other breeds mentioned below they are yellowish red or brownish yellow to yellow in colour:

Shared characteristics:

Heavily built body, ear patches or lobes red, reliable brooding drive, yellow- to brown-shelled eggs.

- Brahma
- Cochin
- German Langshan
- Orpington.

Eggs and Meat

Naturally, humans want to have both eggs and meat. For this reason, farmed poultry include the dual-purpose chickens belonging to the medium-weight breeds. Most of them are cross-breeds of light laying varieties with heavy table birds. These include the Sussex, an old farmyard chicken from southern England. The hens of this breed commonly produce 160 yellow or yellowish-brown eggs in their first year and 130 in their second. They should not be used longer than this in order to utilise their meat at its best. A fully grown cock weighs up to 4 kg and a hen about 3 kg.

Another classic breed of medium-weight chicken with a very high proportion of Cochin blood is the Wyandotte. Its fulsome, soft plumage has inspired breeders to produce many different colouring versions. And for this reason they can be found, like many others, in white, yellow, or cuckoo, silver colours, partridge colours and all the way through to black. In any event, among these dual-purpose breeds, there is a very large selection of birds suitable for small-scale poultry keeping. The most important of these are the following:

Shared characteristics:

Powerful build, red ear patches, mostly yellow legs, often reliable brooding drive, brown-shelled eggs.

- Australorp
- Barnevelder
- Blausperber
- Deutsche Reichshuhn
- Dominique
- Dresdner
- Faverolles
- New Hampshire
- Niederrheiner
- Plymouth Rock
- Rhode Island Red
- Saxonian
- Sulmtaler
- Sundheimer
- Sussex
- Vorwerk
- Welsummer
- Wyandotte.

Brood Hens

The medium-weight breeds have a high proportion of reliable breeding hens among them, whereas among lighter breeds, good hatchers are much less commonly found. However, the reader who has decided for a lightweight breed but nevertheless wishes to hatch and raise chicks in the natural way would be well advised to keep one or two hens of a medium-weight or heavy breed, which can be trusted to care for the offspring.

Good breeders among the medium-weight varieties belong, among others, to the Australorp, Barnevelder, Faverolles, Plymouth Rock, Sundheimer (which are reliable early breeders), Sussex and Vorwerk hens. Among the Wyandottes, the brooding drive varies greatly according to colour type. But if it is present, they breed very reliably; among the partridge-coloured variety, the early onset of broodiness is actually a breed feature.

Among the light laying breeds, as already mentioned, the brooding drive is consistently very weakly expressed; this is the case also with the purely commercial breeds such as Leghorns and the laying hybrids used for egg production and this suits the egg producers well, since broody hens do not lay any eggs.

But we should not expect from our farmyard chickens – and we will return to this at the end of this section – any great achievements in terms of laying and fattening. They are far inferior, from the performance standpoint, to the overbred commercial fowl varieties. Bred, as they are, or rather crossed from pure inbred strains, usually between a laying breed and a meat breed, these so-called hybrids achieve *superchicken* output. For example, within barely 20 years, their laying performance has doubled from about 130 to roughly 260 eggs per year, admittedly exploiting cross-breeding success with a carefully worked-out keeping system and high-quality feeding formulae.

Selection According to Available Space

We will consider the question of space requirements for chickens in the hut and the run in the chapter on housing. At this point, we examine a few fundamental aspects to assist in decision-making.

As a rule of thumb, the larger the chicken, the more space it needs. However, within this very vague range, many factors that do not fit into the weight scheme mentioned above need to be considered. Firstly, the weight and size differences between, say, the light laying breeds and the medium-weight breeds or between these and the heavy breeds are fluid, and secondly, the various breeds have very different temperaments within each category.

It is therefore advisable to gather information on these topics from the relevant specialist literature or from local poultry breeder clubs before deciding on the question of the space available or needed.

For example, Dresdner chickens, which belong to the medium-heavy category and have a good deal of spirit, need the same freedom of movement as Dorkings, which are representatives of the heavy breed category but which have a very quiet temperament. There are also large differences in this regard within any category; one example is among the light layer breeds, where the Krüper with

Good to know

Hens of the heavy varieties all breed reliably and also have the advantage that they can be given more eggs to sit on.

Good to know

In many cases, the temperament of the animals is far more important for their space needs than to which of the three categories they belong.

Helpful tip

For the person who finds it difficult to decide, there remains the possibility of mixing individuals of different breeds and colours and to enjoy a varied flock. It can also be very interesting in the course of one's chicken keeping career to try some experiments and to produce some – often unexpected – chicken varieties.

its short legs and trusting character is certainly content with less space than the Italian breeds, which are considered extremely lively.

The decision is made easier when one considers whether, rather than a large breed, it would be preferable to keep a bantam type or a miniature breed derived from the large fowl varieties. If so, then we can confidently deduct a third or, in some cases, even half of the area in our space calculations. If we assume three to four animals per square metre of hut space for a medium-weight breed, then on the same floor area, at the outside, we could accommodate six to eight animals of the miniature variety of this breed, or we could design the hut space correspondingly smaller for three to four miniature chickens.

Selection According to Personal Taste

Here, it is difficult to offer any basis for deciding, since taste is so individual. Therefore, the only advice that can be given is that the sensible small-scale chicken keeper, if he is not thinking of the animals' expected usefulness, should firstly base his decision on the available space. Then, selecting from the breeds remaining to choose from, he can tend and care for what is to him the most attractive and pleasant type of chicken, with a good conscience.

CHAPTER 3

Anatomy and Physiology of the Chicken

Even the small-scale chicken keeper should have at least a basic knowledge of the form and structure of the body parts and the different vital processes of his animals. This knowledge is important particularly for recognising and treating common chicken ailments, as well as for proper husbandry.

It can surely be assumed about the keeper of a small flock of chickens who devotes a large part of his free time to living creatures, that due to this very interest, he is open to learning about natural life processes.

As we know from school biology, the cell is the smallest living component of a life form, including a chicken. It comprises the cell body, which consists mainly of protein along with many other substances, and the nucleus, which carries the genetic material.

A number of cells amalgamated into one functional unit is known as a tissue and four different types of tissue are distinguished: connective tissue, muscle tissue, epithelial tissue and the blood. The harmonious cooperation of the organs results finally in the living body.

Once the living being is adult, cell multiplication ceases. From this point on, only worn-out and dead cells are replaced by new ones. A vivid example of this is the moult in birds or the shedding of skin in reptiles. Similar processes also take place in internal, and therefore invisible, life processes.

If we wish to appreciate and understand our domestic chicken in its entirety, we can

Some facts and figures about the biology of the chicken	
Body weight (depending on breed)	1.5–4.5 kg
Body temperature	40–43°C
Heart rate	350–480 per minute
Respiration	20–40 per minute
Blood quantity	6.5–8% of body weight
Age of hen at sexual maturity	5½ months
Age of cock at sexual maturity	5 months

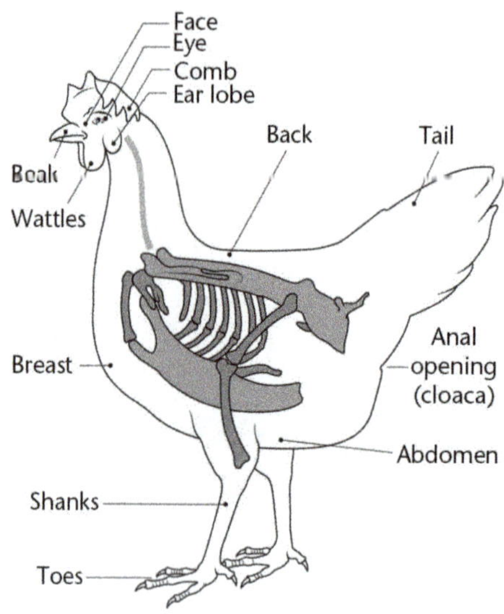

The most important body parts of the common chicken.

begin with the exterior and then consider the more complex structures and processes in the interior of its body.

External Appearance

In conversations between chicken keepers, and particularly among single-breed poultry breeders, the visible body parts of the animals and the specifically developed judging criteria relating to them play a decisive role. To be able to join in with these conversations and to assess our animals from their most important, purely externally recognisable features, the following information is useful.

The Main Body Parts

In order to judge the health and merits, that is to assess an individual animal for its "conformity", we take account of the following:

- the head with face, eyes, beak, comb and wattles,
- the neck,
- the back, wings and tail with associated plumage,
- the breast and belly,
- the hind limbs with thighs, shanks, toes and spurs.

Judging Criteria for a Good Laying Hen

The head should appear fine, narrow and feminine, the beak short and powerful, the face smooth and red. Comb and wattles, which are formed large or small depending on the breed, should be well supplied with blood vessels, the eyes prominent, large and bright.

The breast should be broad, deep and muscular, the abdomen broad and soft.

A healthy cloaca (anal opening) or vent appears moist, pink and without any folds.

Straight, wide-set shanks (legs) and powerful toes with close-lying scales finish off a positive appearance.

Plumage

A close look at a feather can inspire a real sense of wonder at one of nature's truly

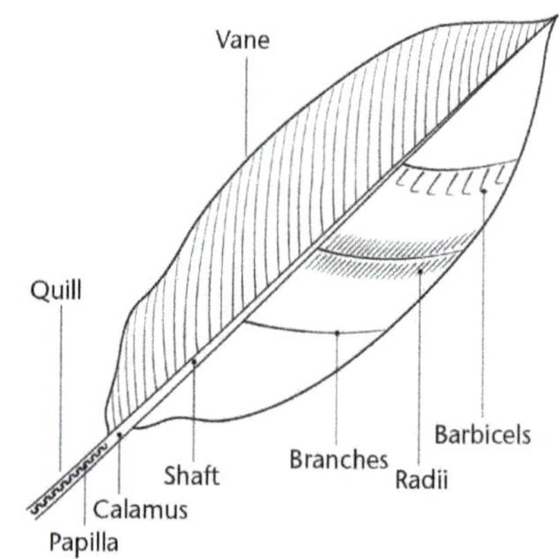

Despite its delicate structure, a feather is able to withstand enormous forces – truly a wonder of nature.

ingenious designs. If we consider what an important set of functions the plumage must fulfil for a chicken, it becomes clear that the intricate and detailed solutions it provides have not come about by chance.

Structure of the Feather

On a first examination, the essential components that are immediately noticeable are the quill and the vane. The quill itself consists of a lower part, the calamus and the significantly longer upper part or rachis. The calamus is hollow, translucent and is anchored, with its navel-like depression, the papilla, in the epidermis of the chicken.

The shaft carries the vane on both sides. The vane itself is made of individual branches, which possess, at their edges, a fine-membered fringe of countless fibrous radii. Finally, arranged at the ends of these radii are microscopically small hooks, or barbicels, with the aid of which they interlock to form the closed vane. At the lower part of the vane, the barbicels are lacking.

As a result, the feather appears loose and down-like at this end.

Feather Types

Depending on the function and structure, a distinction is made between covert feathers and contour feathers, fluff feathers and down feathers, bristle feathers and filoplumes.

Good to know

On cold days, it can be seen in chickens and other birds that to protect themselves against the cold, they puff their plumage up and thereby surround themselves with an insulating jacket of air.

The *covert feathers* form the actual feather covering and therefore the outer extremity of the body of the chicken. They provide protection and a screen against the weather, particularly water, and also against injury. And for this purpose they have a particularly stiff shaft and a sturdily formed vane. The covert feathers on the wings and the tail are known as *remiges* and *rectrices*, respectively.

The *down feathers* are delicate and loose according to their different purpose. Their quill is much thinner and has long branches with fine, thread-like radii on which the barbicels are lacking. Down feathers are found between and beneath the covert feathers and, with their loose structure and the air cushion that they create between them, offer the hen excellent thermal protection. In the absence of down feathers, the incubation of eggs would not be possible. Carefully watching a broody hen preparing to sit, one can observe that she puffs up her plumage and then lowers herself carefully onto the eggs with this air-filled jacket. After all, it is only with this air cushion between the soft down feathers that she is able to establish and maintain the required incubation temperature.

Chicks are sometimes found to have their down stuck together. This phenomenon is caused by excessive thickening of the feather sheath, which has not split off during the drying out of the freshly hatched chick, so that the down feathers also fail to unfold properly. Since this anomaly is a serious problem, these individuals need to be identified and not used for further breeding, if possible.

Finally, the *filoplumes* have a very soft shaft and a reduced vane, which can sometimes be entirely lacking. They can be found between the covert feathers and, above all, at the base of the beak, round the eyes and ears and they appear particularly profusely in some breeds, such as the Silkie.

Feather Formation

The site of origin of the feathers, a body part that is so important for the chicken and all other birds, is the feather papilla, which is visible on the surface of the embryo – in the domestic chicken as early as the end of the first week of development – as a conical elevation.

Between the 12th and 13th day of embryo development, the feather grows out of the papilla to its full size and then, after the 15th to 17th day, as a fully developed downy feather, is enclosed on all sides by the feather sheath. Once the chick has hatched, the feather sheath ruptures and, during further drying, falls off as the familiar chick dust. This releases the soft down feathers that give the chicks the appearance of small loose-feather balls and makes them so delightful to children and adults alike.

Moulting

During its growth and adult life, the chicken passes through several plumage stages. Following hatching, a chick first gains its down plumage. Then it develops its immature plumage and, finally, at about the 18th to 20th week of life, the adult plumage. But that is not the end of the matter. Since the plumage is subject to severe wear, through the influence of the weather or the heavy stresses of flight, it must be renewed from time to time.

Good to know

In any event, the moult involves a significant physical stress for the animals. During this time, they have a sickly appearance, the comb turns pale and they stop laying. The laying organs even reduce in size.

If chickens become too cold, they plump up their feathers for better insulation. The same principle is used by the sitting hen to keep her eggs or chicks warm.

The Moult

The changing of the whole plumage is known as the *moult*. It normally occurs after 12 to 15 months of laying activity as the *full moult*. This is a natural and many-sided physiological process which corresponds to a coat change in a mammal. The onset and duration of the moult vary greatly between individuals, but in our climate, it usually takes place in the late autumn or the winter, although it can be delayed or accelerated by other influences such as variations in the housing or the feed, or a change in the weather. It lasts for two to three months, but busy and consistent layers often have the shortest moult period.

The course of the change of plumage also varies greatly. In some animals, it takes place in stages, whilst others practically get "undressed" in a virtually complete shedding of the plumage for a time.

Apart from the full moult described, there is also the *partial moult* or *neck moult* where the plumage change is restricted to the neck area. In full-grown hens, a neck moult might take place after a strenuous winter of laying activity, although typically their laying time is only interrupted for a short period.

This phenomenon is unwelcome particularly among young hens who have started laying very early but have not been sufficiently well nourished or whose wellbeing has been compromised by being kept in overcrowded huts.

Finally, the *forced moult* should be mentioned, which can be induced by abruptly withholding feed, water and light. The aim and purpose of this measure is to bring about a rapid moult in the whole flock. But since only such a drastic deprivation can produce the hoped-for result, this method is highly controversial.

The Comb

This body part is one of the most noticeable external features of the chicken. Depending on breed, we can distinguish the single comb, the rose comb, the pea comb and the walnut comb. The most commonly found form is the *single comb*, which incidentally is that which graces the main breed type of domestic chicken, the red junglefowl type. It is, after all, from this comb shape that humans have bred the other comb shapes, in that, when mutations occur, they are kept as curiosities and are then incorporated by selective breeding. This simple breeding principle has naturally been used successfully for all the other possible features.

But a hen's comb is not merely an adornment or a breed identifier. The size and the colour are heavily dependent on the influence of hormones, as we can see from the example of a broody hen. Errors in feeding and keeping are also evident from the appearance of the comb. For example, animals kept exclusively in a hut often have large, limp combs. It is suspected that the enlargement of the comb is intended to compensate for a lack of sunlight.

Among specialist breed keepers, a fault-free comb with the properties that have been specifically defined for that particular breed are highly prized. And for this reason, some sneaky individuals have been known to attempt to give nature a helping hand with scissors or a knife. But happily, this kind of animal cruelty is quickly recognised by the trained judges who examine the birds at poultry shows and is severely punished with a showing ban or other sanctions.

Comb Shapes

The *single comb* should have five serrations. Side spikes, various indentations and stunted

As a breed characteristic, the comb shape significantly influences the face of a chicken. Shown here are the four most important comb shapes.

This cock bird with a classic single comb conveys an impression of strength and energy.

growth also often appear in the comb blade, but these slight deviations from the ideal comb do not affect laying performance.

However, if we plan to exhibit our animals as pure-bred birds at shows, we need to make additional breeding selections in this regard.

In the male, the comb should stand upright, while that of the female may flop over to one side. Breeds associated with this comb form include the various colours of Leghorn.

The *rose comb* has a broad base underneath and is covered with small fleshy nodules of even height. It extends towards the neck with a free-standing end, called the "leader". Representatives of this comb shape are the Hamburgh, the Rhinelander and one variety of Rhode Island Red.

The *pea comb* consists of three rows of pearl-shaped lumps arranged in lines, the middle row standing above the others.

Typical representatives of this variant are the Brahma and the Indian Game.

The *walnut comb* consists of an unstructured fleshy knob of variable thickness. Birds with this mostly modest adornment are the Kraienköppe and the Malay.

Notable special comb forms are the *horned comb* and the *antler comb*, which give the French breeds of La Flèche or the Crève-coeur the look of small demons.

The Beak

Once, during the evolution of living beings, nature had converted the front limbs of birds to wings, the beak acquired particular importance as the "tool" for the intake of food and water, nest building and defence.

It has a particular shape depending on the lifestyle of the individual bird type, the beak of birds of prey having a form distinct from that of seed eaters, and the beaks of insectivores being different from those of omnivores, whilst wading birds have bill shapes that are unlike those of land-dwellers, and in woodpeckers, the beak has developed into a highly specialised tool. Beak shapes that are regarded as anomalies among domestic chickens, such as "scissor beak", appear in the

Helpful tip

Since birds with severe beak malformations cannot make optimum use of the run with its rich offering of insects, worms, seeds and green plant matter, it is best to remove them from the flock at an early stage and use them for human consumption.

natural world as a necessary function-related derivation from the norm in the red crossbill. This beak form, in which the upper beak lies across the lower beak and the tip is severely curved downwards, appears quite often as an abnormality in domestic chickens. The reason for this phenomenon is typically an asymmetrical formation of the jaw and nasal bones. Chickens with deformities of this type have difficulty picking up food items. And as a result, their output may be severely restricted or their life may even be jeopardised by it. We can help them by supporting the beak and by providing them with deep and well-filled feeding containers.

Spurs

Spurs on the inside of the shanks are normally a feature of the cock, although older hens sometimes also have them and this is suspected to be due to a slightly altered balance of sex hormones, in a similar way to female facial hair in humans. The spur has a bony core that is surrounded by a spongy tissue and is covered with a horn layer which, like a claw, is constantly worn away and renewed. Long spurs are undesirable in breeding cocks because they represent a risk of injury to hens during mating. In such cases, therefore, the spurs are shortened and rubbed with caustic soda to suppress their re-growth.

Some cocks even have double spurs, which are arranged one immediately above the other on the shank.

The following extract from an old textbook reveals that the spurs, which are peculiar to chickens, provided opportunities for deception:

As is known, the age of a cock can be judged fairly precisely from the spurs, but since chicken dealers have diverse similarities to horse traders, it has been known to occur that a very old cock has been made seemingly younger by twisting off some of the years-worth of spur growth with a pair of pliers.

The spurs play a part in some countries in an activity that is regarded as despicable by Westerners, specifically cock fighting, which enjoys a similar tradition there to the bloody entertainment of bullfighting. The fighting cocks are specially bred for this "sport". For some fighting-cock owners or fight organisers, it is often not sufficient for the animals to cause each other serious injuries with their sharp spurs. To make matters worse, they also tie small razor-sharp knives to the spurs, since this, for the spectators and those laying bets – as money always underlies such activities – heightens the bloodthirsty enjoyment.

> **Further info**
>
> *Whereas the archaeopteryx still had teeth, claws, relatively heavy bones and a primitive brain, our modern bird species are highly specialised and intelligent animals compared with reptiles.*

The Skeleton

The Evolutionary Perspective

It is now known that birds evolved from reptiles. This is shown by the great similarity of the skeleton of birds to that of the scale-covered reptiles. Similarly to the archosaurs, during the course of their evolutionary history birds developed the ability to walk on their hind legs. They also acquired from their dinosaur ancestors the ability to fly, since even before the precursor of the birds, the archaeopteryx, a pterosaur with a wingspan similar to a glider, populated the earth. Only later did flight-capable lifeforms appear which, instead of flight membranes, had feathered wings and eventually took a different course of development from the reptiles.

For a long time, humans were left watching longingly as the birds soared high above in the sky. But flight had only been made possible by

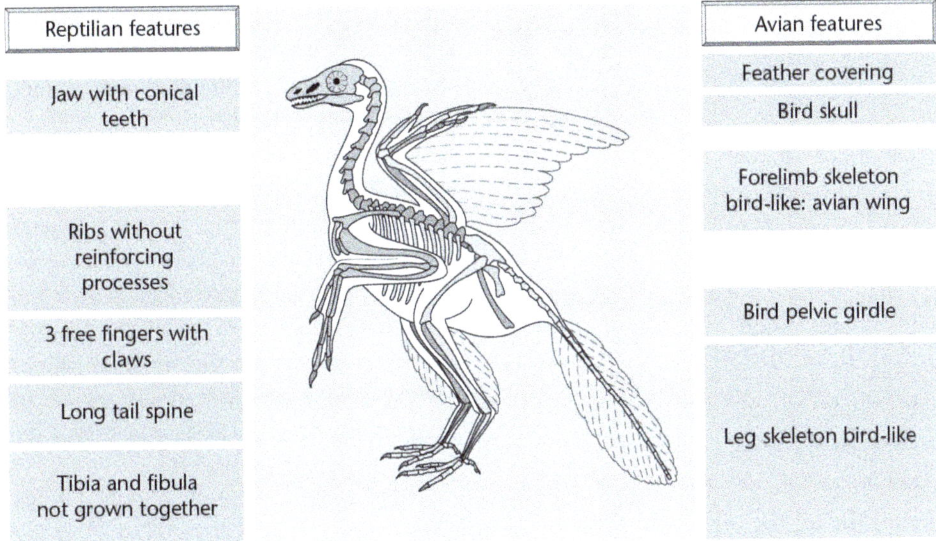

The primitive bird archaeopteryx unifies, in its body, features both of land animals and of the birds that would evolve from it.

millions of years' worth of step-by-step adaptation of the skeleton to the demands of this new lifestyle and means of locomotion.

The body became shortened, the long tail of the reptiles shrank, the skull grew compacter with simultaneous enhancement of the contents – namely the intelligence – and the bones acquired air-filled hollow spaces to lessen the weight. Among the bird species living today, there are highly varied body shapes, adapted to their lifestyle and feeding habits. A typical cursorial bird, the European bustard, has a very different structure to a swift, which spends the greatest part of its life in the air and has stunted feet. Ducks have a specialised bill and webs between their toes for propulsion in the water. The domestic chicken is certainly able to fly like most other land fowl, but mostly it walks and scratches about on the ground, rising into the air only for short distances in an emergency. Its body shape and particularly its powerful legs and feet are well adapted and suited to this lifestyle.

The Skull

The individual skull bones and plates form the characteristic head of the chicken with its laterally positioned eyes, almost invisible ears and pointed beak.

The head contains and protects the brain and the sensory organs. According to these different functions, zoologists differentiate the cranium and the facial skull, the extent of the latter being essentially defined by the form and size of the beak which, in seed-eaters like the chicken, or in water fowl like the duck, makes up the larger part of the head.

The central nervous system – i.e. the brain – is housed in the *cranium*, which has a shape like a large curved plate.

The *facial skull* with the beak as the defining element is the container for the sensory organs, such as the eyes which are set in the sides of the head and are well padded with a fat layer in the skull cavities provided for them.

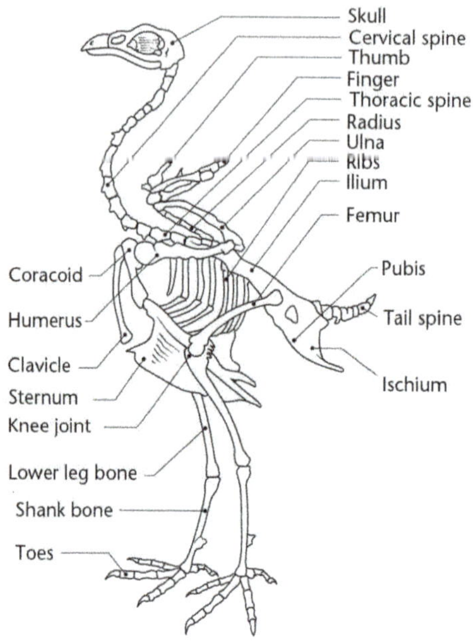

The skeleton of the domestic chicken showing the most important bones.

Torso Skeleton

Moving down from the head, firstly, there are the 13 cervical vertebrae which form the S-shaped neck spine. With this arrangement, the chicken is easily able to balance its head and carry out head movements in all directions. This is important not only for speedy recognition of enemies, but also for proper plumage care, since birds use their beak for this purpose. In addition, the first vertebra, the *atlas*, is somewhat smaller and is shaped so that the chicken is able to rotate its head through 180°. This phenomenon does not apply only to chickens, of course. Owls hold the record in this respect and just watching them is enough to make one feel dizzy.

The seven thoracic vertebrae that follow are largely fused by bony growth to one another and to the pelvic portion of the spinal column. The 13 or 14 lumbar and pelvic vertebrae are also fused with one another and form the actual pelvic girdle. The end of the spinal column is formed by six caudal, or tail, vertebrae, the last of which is large and flattened and so offers the tail musculature an excellent attachment site to perform its steering function.

Matching the number of thoracic vertebrae, the chicken has seven pairs of ribs which, together with the strong breast bone, or sternum, form the rib cage which protects the internal organs and intestines. The sternum is the largest bone in the skeleton and resembles a mighty shield with a blade or keel on the underside, positioned exactly in the middle. The latter is an excellent attachment point for the powerful flight muscles. However, curvatures and deformities of this important bone often occur. Some of these are genetically caused and some come from faulty nutrition (for example, a deficiency of calcium or vitamin D) or arise from imperfect husbandry. It should be avoided, for example, that the young chickens get used to using perches too early – that is, during their growth phase – since the frequent jumping up and down again leads to such deformations in the breast bone.

The part of the pelvis where, in hens, the ovary and the oviduct are situated is formed by the fused-together ilium, ischium and pubis. With a little practice, we will be able to discover which hens will probably make the best layers by a testing feel between the pelvic bones. When performing this test, however, it is important to proceed gently.

Good to know

The distance between the two hind-most "bones", also referred to as the pubic or pelvic bones, gives a relatively reliable indication of the laying performance of a hen.

Limb Bones

Remaining in the region of the pelvis, it can be seen that the thigh bone is attached in the articular depression of the ilium in a way that is not visible from the outside and is joined by the knee joint to the lower leg. The tarsal and metatarsal bones are fused together into a powerful tarsometatarsus, or shank, and this is often wrongly thought of as the lower leg bone. This impression is also primarily reinforced in a disinterested observer due to the fact that the thigh is not visible and the knee is also covered by plumage. The shank is then connected by joints to the toes. Most chicken breeds have four toes. Exceptions are frizzled chickens and Silkies, which nature has blessed with an extra toe. Since the toes can be curled without the need for special muscles, a chicken can perch for long periods awake or asleep on a branch or a pole without falling off.

The front limbs arc adapted for the alternative locomotion method of the chicken, flying. They are attached to the bird's body by means of the pectoral girdle, which consists of a long narrow shoulder blade lying close to the sternum, the clavicle or collar bone, and the powerfully developed coracoid bone. All three of these bones are united at their joining point in an articular depression which provides a rotation and pivot point for the wing. The wing itself can be regarded as an arm adapted for a special function, with a powerful upper arm bone, a lower arm made up from a radius and ulna and a vestigial hand with three stunted fingers including the thumb, which has a small claw. All in all, the physical structure of the chicken as described represents a harmonious whole which is well suited to its lifestyle and feeding habits, altered over millions of years through accidental changes to the genetic material, initially supported, and finally optimised, by the external environmental influences acting on it. If we compare it to the physical structure of a cat with its long, flexible spine, short neck, long tail and generally low, elongated physique, we find the same degree of harmony, but adapted to a completely different lifestyle.

But we will not end our consideration of the inner life of the chicken there. In the following section, we will look at some more remarkable facts about our little domestic bird.

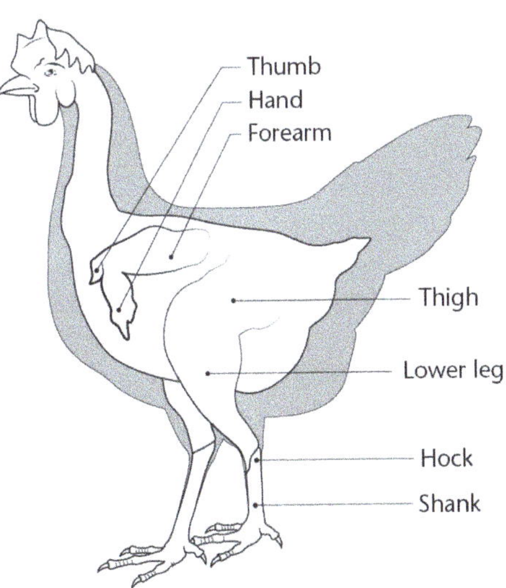

The position of the limbs is made clear if – as here – they are shown embedded in the outlines of the body.

The Senses

In order to understand how chickens react to us and their environment, we need to know

Good to know

The chicken, like most birds, has very sharp vision which is developed for the rapid identification of nearby objects.

how well developed their senses are, and so whether they react more strongly to visual stimuli than to sounds, whether they are able to perceive smells and how acute their senses of taste and touch are. Learning a little about these particular characteristics can prove to be quite a revelation.

Vision

Whilst good fliers like doves, geese and ducks or plain-dwelling birds like the turkey must be able to survey a wide area to recognise suitable food items or enemies in good time, the chicken, as an inhabitant originally of jungle and shrubland, relies on having sharp vision for objects in its immediate vicinity.

For this reason, chickens therefore hardly pay attention to things happening at a distance, that is, more than about 50 metres away.

Their three-dimensional vision is restricted to the region in which the fields of view of the two laterally arranged eyes overlap. The consequent lack of depth perception can be replaced to a certain extent by alternately fixing an object with the left and right eye through turning the head back and forth or by approaching the object of interest in a zigzag fashion.

Even an enemy, in the form of a hawk, for example, can only be made out by tilting the head. An inexperienced observer might find this head position comical, but for the chicken it can be literally a matter of survival or not.

One might wonder also what the chicken's colour vision is like. The fact is that chickens can recognise colours clearly. This has been shown by scientific experiments, although it can be strongly influenced by different brightness levels.

Chickens are very inquisitive and sometimes also playful. An old plastic bucket makes a wonderful object for training all the senses.

Good to know

It has been found, in general, that adult chickens and chicks alike are more attracted by familiar sounds than by familiar sights. Try it out for yourself. You will find that even animals 50 metres away react to familiar enticing calls that promise fresh food and water.

The Structure of the Eye

The eye of the chicken is essentially structured the same as in mammals. The eyeball is set in a cushion of fat, protected against pressure and shocks. The differences lie only in the shape of the eyeball, which is more disk shaped, and in the more strongly convex cornea, which is additionally reinforced by bony deposits. A comb-like structure, the pecten, extends from the site of entry of the optic nerve into the vitreous body situated behind the lens. It is believed to increase the sensitivity of the eye to moving objects. It is important, above all, for recognising and catching flying insects and small creatures in and under the grass cover, leaves and shrubs.

A further differentiating detail is that the chicken eye has, in addition to the upper and lower eyelids, a third lid, the nictitating membrane, which can be drawn from the inner to the outer corner of the eye to cover it. It has a protective and cleaning function.

Hearing

The ear of the domestic chicken is almost as effective as that of a dog. This capability is very important for a shrubland dweller with only a very restricted field of vision. After all, enemies are more readily recognised in such environments by sounds as they come into the field of view of the prey.

From its 18th day of development, the chick can already perceive sounds inside the egg and is also imprinted with the voice of its mother. Later, when the mother leads her chicks, they react to her voice from a distance of about 15 metres, while the mother can discern the distress calls of her young from up to 20 metres away.

The Structure of the Ear

Compared with mammals, the outer ear or earlobe is lacking in birds, including chickens. A fringe of skin, which is densely covered with feathers, protects the entry into the short external auditory canal leading to the eardrum. This is also convex in form and is stretched across a ring of bone.

The vibrations impinging upon the eardrum are conducted by means of a simple ear bone to the inner ear through its fluid. This fluid bathes the "labyrinth", which consists of the semi-circular canals and the cochlea and represents the actual sensory organ for both hearing and balance.

Taste

The sense of taste is only very poorly developed in the domestic chicken, like other seed-eaters. Nevertheless, it can distinguish very well between the four basic qualities of flavour: salty, sweet, sour and bitter. The taste buds for these are situated in the beak cavity, under the tongue, in the pharynx and in the throat. Due to the low taste sensitivity, these four taste variants play only a subordinate role in the chicken's choice of food.

Food selection is based more on the graininess of a food item and on its external qualities, such as hardness, softness or rawness. The chicken also proves to be very insensitive to bitter-tasting substances, whereas it strongly rejects sour-flavoured food items.

Touch

Far more important for the animals in their choice of food is the sense of touch. Numerous touch receptors are distributed over the beak cavity, on the tongue and at the tongue's margin, as well as in the pharynx cavity, and these convey impressions to the animal regarding the size, hardness and surface quality of the food. Thus a bird learns to recognise the shape of wholesome feedstuffs. Based on experience, over time it becomes easy for it to distinguish wholesome from unwholesome items. Our domestic chickens also gain a more or less great wealth of experience in this regard, depending on how they are kept.

The touch receptors are formed in the shape of small pressure pads made of liquid-filled cells. If, for example, a seed makes contact with this pressure pad, the stimulus due to the change in shape of the cells is transmitted to the adjacent nerve tissue and from there is communicated to the central nervous system.

Smell

Due to its low importance for the chicken, smell is the least well developed of all the senses. In fact, the domestic chicken is considered to have an extremely poor sense of smell, and this is apparent, among other things, in that they drink slurry, peck at rotten eggs and happily scratch about in animal dung.

But this does not mean that we should not be particular with the drinking water and the air in the hut. In the wild, junglefowl would hardly have an opportunity to drink slurry or to consume dung in great quantities. The air is also incomparably better. We should always remember that contaminated feed and water, as well as foul air, can make our animals sick, not just because it stinks, but because dangerous microbes exist and find ideal growth conditions there.

Breathing

The breathing of birds differs from that of mammals. Whereas in mammals, the exchange of used and new air takes place in the alveoli of the lungs, in birds, the unused air that is drawn in is moved through the lungs into the air sacs and is transported via branches that reach throughout the bird's body, including into air-containing "pneumatised" bones. During the transport back from the remote parts of the body through the lungs, the air releases oxygen again. By virtue of this flow-through principle in conjunction with the air sacs, significantly more oxygen and, as a result, more energy is made available to the bird's body and thus to the metabolic processes for its vital functions.

The body temperature, at 40–43°C, is also correspondingly higher than in mammals, which can develop a body temperature of slightly above 41°C during a high fever; this, for them, is at the limit of survivability.

This highly efficient breathing system is adapted particularly to the highly energy-consuming demands of flying. The hollow bones, the air sacs and the light insulating plumage, which renders an insulating fat layer unnecessary, make the avian body light in weight.

Directly related to parts of this system is the heat-regulating capability of birds. They do not have any sweat glands under their plumage to be able to remove excess heat and, above all, to be able to excrete moisture for the purpose of cooling. Sweat glands like those of mammals would glue the plumage together and render it unusable. And in fact it is the air sacs that assume the function of sweat glands. A large part of the water vapour that otherwise would be evaporated as sweat is collected in the air sacs and conveyed to the exterior along the airways and through the beak to the outside.

And so we should not be overly concerned on hot summer days if our chickens present a

Helpful tip

We can help our birds on hot days by providing them with a well-ventilated, but draught-free henhouse and a run with shade-giving trees and bushes.

Good to know

New types of feed, or the same feed provided in a different form, arouses firstly suspicion, followed by curiosity and, if liked, after a period of familiarisation, enjoyment.

horrifying sight, sitting still with their plumage puffed up, their wings spread out, their beaks open, while panting. They are merely trying, by this means, to cool themselves down.

Digestion

Continuing our comparison with mammals, the differences in birds begin with the first part of the digestive apparatus, namely the mouth. A chicken has no teeth with which to reduce the size of food. And so food passes from the mouth cavity directly – apart from the addition of a certain amount of mucus – into the crop. This is a protrusion of the oesophagus and is constructed to enable chickens to consume and store large amounts of food without interrupting their digestion. It is here that grains can be a little softened and so prepared for the subsequent digestion. From here, the food contained is transferred in irregular waves into the stomach. If necessary, the crop can store a whole day's food intake. If the crop and the stomach are empty, the food passes through the oesophagus, past the crop and directly into the digestive tract.

Glandular Stomach and Gizzard

The *glandular stomach* or *proventriculus* serves to produce digestive juices which mix with the food to form a slimy mass. It is comparable with the main stomach of mammals. Here, the digestive fluids important for protein digestion are secreted by the glands. The actual digestion – that is, the chemical and mechanical conversion of foods into usable components – takes place in the mechanical stomach or *gizzard*.

The gizzard consists of two pairs of muscles, specifically two thin-walled intermediate muscles and two thick-walled main muscles.

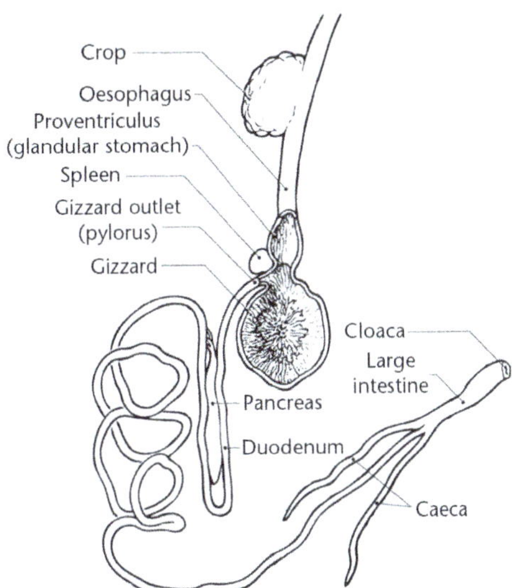

The digestive organs of the domestic chicken shown as a functional unit.

Both pairs contract alternately and thereby create a reciprocating frictional force on the food mass.

Inside it are small stones that the chicken has taken in with its food, and which support the gizzard in grinding down the food mass. This is rather as if we had taken a leather bag filled with soaked cereal grains in our hands and then had to convert the grains into a very fine pulp through kneading motions. If we were to add small pebbles, this task would certainly be made easier. In the case of housing without a run or with a run that offers the animals little opportunity to take in small

Ratio of body length to gut length in different domestic animals	
Chicken	1:8
Duck	1:10
Rabbit	1:13
Horse	1:15
Pig	1:25
Bovine	1:30

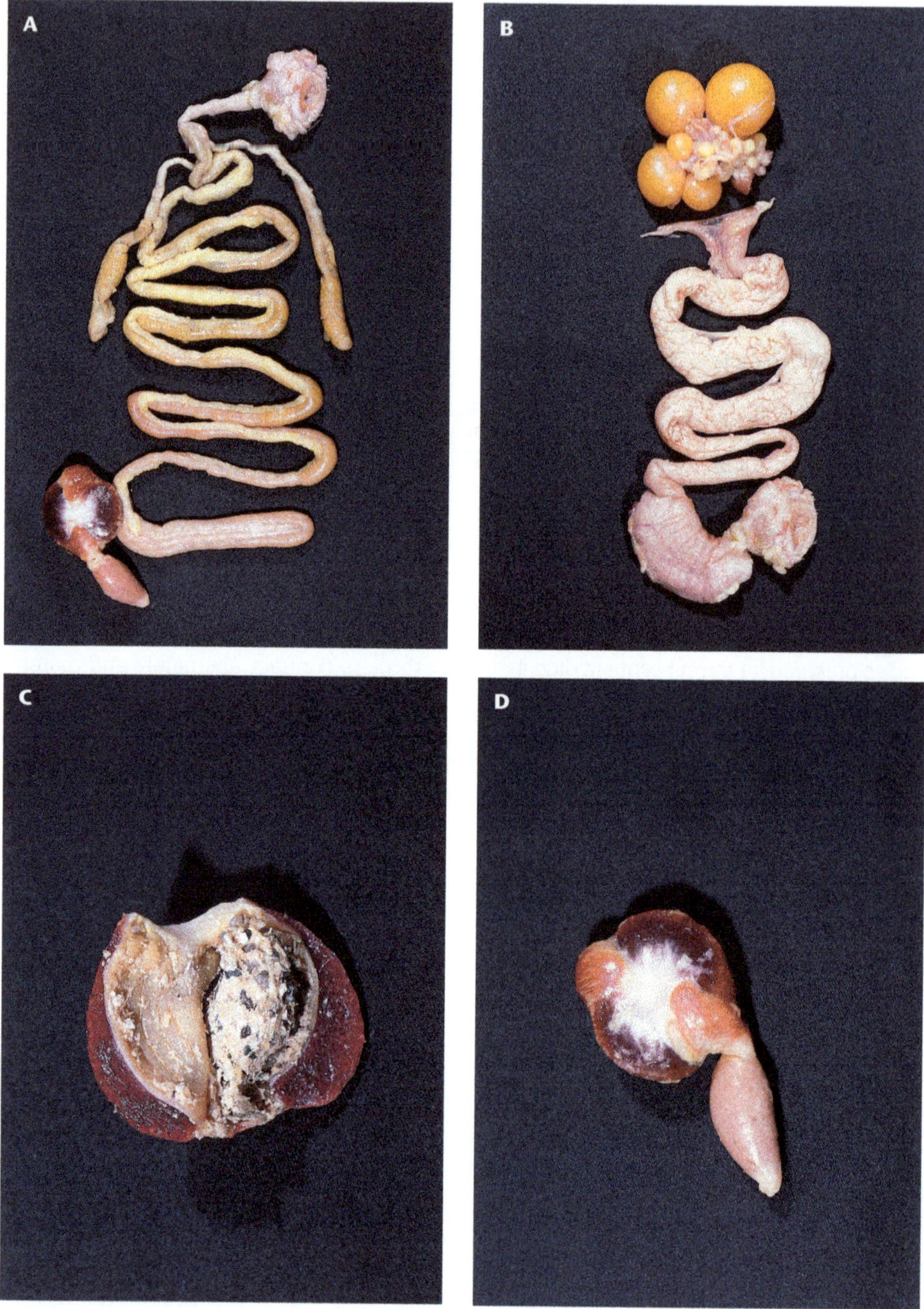

A: The gastrointestinal tract of the chicken consists of the proventriculus and the gizzard, the small intestine and large intestine and the cloaca.

B: Ovary with ovarian follicles of varying degrees of maturity, infundibula, oviduct and cloaca.

C: A section through the gizzard clearly shows the grit that is so important for digestion.

D: Proventriculus and gizzard of the domestic chicken.

stones, we need to provide them in the form of grit. Otherwise, the physiological utilisation of the food would be much less effective, so that the chicken could not draw as much energy from the food as the effective nutrient content of the food actually allows.

The Intestinal Tract

Once the food has been subjected to this intensive treatment in the gizzard for about two hours, it is transferred to the intestines, initially the duodenum as part of the small intestine. The small intestine is equipped with many intestinal villi which increase the size of the internal intestinal surface and offer more sites of action for the enzymes, which break down the nutrients chemically.

The most important source of these substances and therefore one of the most important parts of the digestive apparatus is the pancreas, which is found in a bend of the duodenum.

It secretes:

- *trypsin* to break down protein,
- *lipase* to decompose fat, and
- *diastase* to digest carbohydrates.

The villi then absorb the broken-down nutrients and conduct them in the blood to the individual organs.

The large intestine is significantly thicker than the small intestine. It has fewer villi and ends at the cloaca. Lying between these two intestine portions are the two caeca, which have different lengths. These are effectively fermentation chambers in which the raw fibres of vegetable foodstuffs are broken down and made at least partially digestible to the animal. Raw fibre includes chaff and the husks of cereal grains, the fibrous constituents of their pasture and the kitchen waste of plant origin.

Good to know

Two types of excrement can be distinguished: large intestinal excrement and caecal excrement. One evacuation of the caeca takes place for about every ten evacuations of the large intestine.

The constituents of the food that are still not usable are removed through the cloaca. It is recognisable from the dark brown colour and a foul smell. In comparison with mammals, the whole digestive system is very short. Whilst the alimentary canal of the domestic chicken is only about eight times the body length, the gut of domestic cattle is more like 30 times body length. The digestion times of the chicken are correspondingly short, and the foods containing relatively more water require less digestion time than dry food such as grains or feed meal.

This might explain why chickens are known for being early risers. After all, hunger is the greatest driver of all the activities of life.

Urine excretion

The excretory organs of the chicken are reduced, as compared with mammals, to the kidneys and the urinary ducts, which conduct the urine to the cloaca. The renal pelvis, bladder and urethra are all lacking and the urine itself consists mainly of uric acid. It also contains ammonia, amino acids and various salts. It is removed from the body through the cloaca with the faeces as a semi-solid whitish mass.

Reproduction

Female Sexual Organs

The female sexual organs are formed, as distinct from mammals, only on the left-hand side of the bird's body. They consist essentially of the ovary and the oviduct. The ovary, resembling as it does a bunch of grapes, contains innumerable eggs of different sizes from the microscopic egg cell to a large, yolk-filled ripe egg which may leave the ovary at any moment and can glide into the infundibulum to be encased inside the oviduct with albumin and then with the egg membranes and the shell and so to be transformed into a complete chicken egg.

The oviduct is a tubular organ with a variety of sections geared to egg production, each having a special function of its own, and ending at the cloaca. See the chapter on the egg for more detail.

Male Sexual Organs

The male sexual organs consist of the testes and the spermatic ducts extending from these to the cloaca. The male chicken lacks a penis-like reproductive organ, in contrast to the males of the duck and the swan. The testes are situated at the ends of the two renal poles and the right testis is smaller than the left testis. During the mating season, both testes increase to many times their normal size. They are the site where sperm are formed and stored so that they can be conducted through the spermatic ducts into the cloaca.

The mating act, when the cock climbs onto the crouching hen, takes place in such a way that both birds press their protruding cloacae firmly together by turning their tails aside. The mating itself lasts – as distinct from, say, dogs – only a few moments and can be repeated several times in a day.

CHAPTER 4

Behaviour of the Domestic Chicken

A mere knowledge of the anatomical structure of the chicken and of the life processes that take place in its body do not, of course, afford us anything like a complete enough picture to be able to understand its needs. It is important to know to what, and how, our feathered friends react, how their relationships normally work and how they behave toward other types of animal and toward humans.

The study of behaviour, or ethology, has grown in importance in the last few years in relation to chickens, particularly as a result of increasing concerns about the caging of laying hens. The famous zoologist Konrad Lorenz, who carried out detailed research on animal behaviour, made observations of the domestic chicken early in his studies and reached some interesting conclusions. And since Lorenz, biologists have conducted many more studies, always seeking to discover the true nature of the chicken and with the aim of developing objective principles for their keeping, as well suited to the species as possible. Despite this, a satisfactory solution for large-scale poultry farming has still not been found, at least from the viewpoint of those whose concern is animal welfare.

It is much easier for us as keepers of small flocks of chickens, without the need to make a profit, to create living conditions for our more manageable numbers of animals which meet their differing environmental needs, based as far as possible on prevailing scientific knowledge. And with this motivation in mind, the most important aspects of behaviour will now be outlined, relating these to the real-life situations of the domestic chicken.

Good to know

A precondition for a well-functioning community life is the recognition of a particular hierarchical structure. This is firmly defined by social rank or "pecking order" in the flock.

Life in the Flock

The domestic chicken feels most at home in community. By its nature and its life structure, it is a herd animal and needs the group for survival in its natural environment. The flock offers protection from dangers and a sense of security. But for the advantages it gains, the chicken must obey the rules of the community; it must conform and often accept a subordinate role. Each animal has its precisely defined social position within the pecking order, which it attains after a phase of conflict during the transition from immaturity to adulthood, and then maintains mostly unaltered from that point on.

Hens

As puberty starts, the playful quarrels of the chicks begin to turn into fights (in hens from 10 to 12 weeks and, in cockerels, from 12 to 16 weeks), which can sometimes be bloody. Only at full sexual maturity is a clear pecking order firmly established. If this were not the case, constant wrangling and even violent conflicts would keep the flock in continual flux and would soon use up the energy of the animals. Knowing their social status, individuals go out of their way to avoid trouble. Respect, recognition and a certain distance therefore mark the social conformation of the flock. But things are not always so peaceful. Envy and resentment are the drivers of the occasional squabble and the trigger, for there is usually competition for the best feeding or sleeping place. In these cases, the bird with the highest status within the flock usually has priority over all the others. The best nest, for example, will usually be claimed by the most senior individual first of all. However, in such instances, it can occur that lower-status animals, suddenly empowered by eagerness, refuse to observe the pecking order and actively revolt. On these occasions, the higher-status bird usually holds onto its position, defending itself against the rebel with blows from its wings and by pecking back at it. In most cases, such quarrels are only brief and, with few exceptions, promptly ended, with the higher-status hen emerging as the victor. It can also happen that hens higher up the pecking order victimise certain fellow flock members, apparently just because they don't like them.

The status that a hen achieves among her kind depends on various factors, such as whether she has a bold, self-confident manner and the readiness to fight and persevere. An additional factor is physical appearance, where a large comb and a powerful stature both earn her respect. Finally, age plays a decisive part and, in general, young hens are all lower in the pecking order than even the weakest old hen.

Chickens have only a limited memory for individual relationships, so that too much reliance on this would only lead to uncertainty and thus always to new unrest and bickering. Each hen assesses others based mainly on the shape and size of their comb and on the glossiness and size of their eyes.

Every change within the flock – for example, the removal or addition of a hen – can easily disturb the social order, once established. When introducing strange animals or those that have been kept separate from the flock for a long time, it is best to place them

Good to know

The social order, once worked out as a relatively peaceful community, is only sustainable if the flock remains as constant as possible and does not exceed about 40 birds.

Good to know

In all cases, it is useful to include the cock in the selection process for the group from the start. If a fixed social order has formed among the hens, it will be difficult for him to fit into this conspiratorial community.

unobtrusively on the perching rods among the sleeping hens under cover of darkness.

By this method, it can at least be avoided that a newcomer is mercilessly pursued by the whole group. However, it often takes weeks until complete integration or re-integration into the flock is completed.

Chickens get along well with most other domestic animals. Here, a little glimpse of paradise.

Cocks

The cock has a special place in the social hierarchy of a flock of chickens. Once he has asserted himself due to his physical superiority, he then becomes the undisputed ruler – literally the cock of the roost. And in this elevated position, he has an important function, specifically the maintenance of peace and calm among his quarrelsome females. There is a very noticeable difference between a group of chickens with or without a cock, as far as social harmony is concerned. And so, he should not be seen as a superfluous mouth to feed or purely as an adornment in the chicken run; he is – provided one's neighbours will tolerate his loud crowing – a good investment in any event. If cockerels grow up under the sovereignty of a fully grown cock, his dominant status will remain unchallenged, even if they become superior to the older cock in strength and size.

We have ourselves witnessed, in our ten-strong firmly established flock, that a young

As a conscientious defender and guardian of his flock, the cock has an important role in chicken society.

The cock will aggressively defend his hens against enemies, in addition to keeping peace within the flock.

cock of about the same weight class had to take flight faced with the attacks of the furious flock.

In another of our attempts, a heavier older animal of impressively masculine appearance had a struggle to become master over the females. Particularly the highest-status hen did not wish to give up her position of power and defended herself vehemently. Finally, the heavy cock won out and from then on maintained his ruling position without challenge.

Cock and Hen

Chickens are polygamous, making them different from geese, which live in bonded pairs. The harem of a cock includes about 10 to 15 hens. If the flock is any larger, the danger exists that not all the eggs will be fertilised. Although the cock certainly does not distribute his favour equally, it is seldom that a hen remains entirely overlooked by him. But, true to his lord-and-ruler role, he keeps one or more favourites whom he particularly courts and with whom he more often mates.

Courtship

Before each mating, the initiative is taken by the male party. The cock tries to draw the selected hen's attention to him with, often only feigned, titbits or a particularly attractive nesting place to entice her, or he attempts to approach her directly. During food-offering, he pauses either in a bent, searching posture or stands upright with a titbit in his beak.

Some less gallant males sometimes themselves swallow the delicacy they offer, but a skilful wooer gives it kindly to the object of his attention.

If the cock resorts to the attractive nest tactic, or "nest-enticement", he crouches down in as cosy a corner of the hut as he can find, scrapes out a flat hollow and, uttering cooing sounds, lures the hen he has his eye on to come over to it.

A special form of homage involves a ceremony of "stumbling over a wing". The proud male circles the hen with dainty tripping steps and stumbles again and again over his downwardly spread wing. If the hen is impressed

Not to be mistaken is the sound of the strutting cock when he lets rip and crows at the top of his voice.

Copulation lasts only seconds with chickens; it is no easy feat for the cock to keep his balance in this position.

and finally crouches receptively, he quickly jumps onto her from behind and performs the act of copulation. However, if she tries to escape, the cock will follow her in the "turkey" position, pursuing her with outstretched neck, ruffled feathers and shuffling wings. If he catches the fleeing hen, he jumps onto her without any further ceremony.

Mating

During a typical mating act, the hen assumes a slightly crouched position. The cock approaches laterally from behind and climbs onto the willing hen with outspread wings to keep his balance. At the same time, he grasps the neck feathers of the hen in his beak and clamps her wings with his feet.

Finally, by turning their tails aside, both animals press their cloacae firmly together. At this point, the sperm emerging from the seminal canal of the cock are transferred into the hen's body. Once copulation is complete, the cock dismounts and performs another brief courtship display by way of an encore. The hen is no longer impressed by this; she merely shakes her feathers and carries on with the rest of the day's business.

Egg-laying

In selecting the nest to which the hen will entrust her egg, she is faced with a weighty decision every time. Not every nest offered is accepted and often she seeks out her own place and arranges it accordingly.

Nest Selection

As a scrubland dweller in its natural habitat, the chicken prefers somewhat secluded sites in half-shadow, if possible, situated slightly elevated and with soft clean nesting material. We can meet this need if we structure the nests as described in the section on equipping the henhouse. Our own hens, for example, seem to like following the example of their predecessors.

Thus we observe that even used nests are preferred, especially if they already have eggs in them. If a nest is besieged by several hens at once, it is always the higher-ranking individual who has first choice. However, it can also occur that several animals co-exist peacefully in one small nest. Based on this observation, some chicken keepers offer "family nests" instead of, or as well as, individual nests.

It is interesting to see how the animals behave before laying. The intensity of nest-seeking is not only different between breeds, but is also very strongly dependent on the respective temperament of the individual. Most animals isolate themselves from their fellows during this phase. They become restless and

Good to know

So as to save ourselves a lot of trouble and searching around, we should take care to create attractive nests for our hens.

pace about, clucking. Then again, some hens behave in a very agitated manner and sometimes lapse into more hectic behaviour, even jumping up at the hut wall for minutes at a time.

Laying

Before a hen settles down for egg-laying, she first inspects several nests, although she will usually decide on the same one that she had chosen for the previously laid egg. Some animals can spend hours in a quandary over the decision, but once a hen has made her choice, she plucks the nest litter into shape and deepens the middle a little and, like a dog, turns round and round before settling down. Finally, she gets down to it and deposits the egg in a slight squatting position.

Post-laying Behaviour

Once the egg has been laid, individual hens behave in very different ways. Whilst some remain a while quietly in the nest, as though exhausted or relaxed, others rush out and announce their deed with a loud cackling sound. This has been construed in various ways, but scientists theorise that among the wild-living forebears of our domestic chickens, the post-laying clucking is intended to restore the connection with the now more spread-out flock. Thus, this utterance is also regarded among animal behaviourists as a "herd-seeking call". The cock, in particular, is supposed to be sensitive to this call of the hens and will rush toward the absent, calling individual to guide her back to the flock.

Nesting and Brooding Habits

A hen that is becoming broody is easily recognised by her sharply altered behaviour in relation to her fellow chickens. She increasingly separates herself and noticeably avoids contact with other hens. If another flock member comes too close, she will try to avoid it with hasty fleeing and simultaneous attack-warning movements. If the cock tries to woo her, she keeps him away with plumped-up feathers and a rejecting demeanour.

The broody hen builds – if at all – only a makeshift nest. As a general rule, she will seek out a ready-made nest or a protected, softly padded site and will simply create a flat hollow by sliding back and forth. She lays the eggs in this and as soon as she is satisfied with the number of eggs, she begins the incubation process. Once the hen is firmly established on her eggs, her behaviour becomes calmer again. In general, she only leaves the nest once a day to feed and to perform other bodily functions. If, however, fellow hens or other animals come too near her brood, she reacts very aggressively.

Good to know

The hen is not so jealous as to want only to incubate the eggs she has laid herself. She will also accept nests that are already occupied by a number of eggs.

And it is exactly this behaviour that makes it easy for us to have eggs incubated by animals that produce good broods.

In their first days of life, the mother hen keeps the chicks close together. She also keeps a prudent distance from the other flock members, so that her little ones will not be attacked.

Mother–chick Relationship

Even in the egg – from about the 17th day of incubation – the chicks perceive sounds from their environment. Naturally, what they hear most often is the clucking sounds made by their mother. It is not surprising then that, after hatching, they are so well imprinted with these sounds that they recognise their own mother from among others by her voice. Conversely, the quiet cheeping sounds that the – as yet unhatched – chicks emit from the egg are also of great significance for the establishment of a close relationship. Experiments have shown that deaf hens who cannot hear the sounds from the egg peck at their freshly hatched chicks and even kill them due to the obviously lacking social bond which normally develops during incubation.

Hatching

If the broody hen feels that the chicks are starting to leave their confined brittle cages, she stays quite still in order to avoid disturbing the hatching process. She also remains quite passive; that is, it would not enter her head to assist a chick that had become stuck to free itself from the shell.

On the other hand, one would never observe a broody hen hindering the hatching of her chicks.

Once the chicks have freed themselves from their narrow prison, their mother moves about very carefully. She barely gets up, let alone moves to leave the nest, so as to avoid injuring the tiny newly hatched chicks.

It is interesting to consider just what a heating effect the hen must generate, firstly to dry out her offspring following their hatching and then to keep them warm day and night. More information about this can be found in the chapter headed Incubation.

Imprinting

In the first 36 hours of their life outside the egg, the small chicks are particularly receptive to environmental stimuli. We therefore speak

The guiding mother hen observes her surroundings carefully. Woe betide anyone or anything that comes too close to her! With puffed-out neck feathers and upright stance, she will protect her brood against any threat.

of an "imprinting phase" in which they are sensitive to the impressions with which they are bombarded.

It is clearly of great significance that these first impressions are indelibly imprinted on the chicks. That, of course, is how Mother Nature orders things, and with good reason.

Chicken chicks are nidifugous: they are able to leave their nest after a relatively short time to explore their surroundings. Now, a mother hen who has perhaps 10 to 15 chicks to look after would have difficulty keeping sight of them all, leading them on the search for food and being able to protect them against danger if they were all to leave the nest at will. This close bond mentioned above is formed through a dialogue that is begun between the mother hen and her chicks when they are still inside their eggs. Added to this, following hatching, is the visual appearance of the mother and the siblings, all of which fixes the little flock into the necessary social cohesion in the first weeks of life. Everything that moves seems particularly to attract the little creatures. So, after a short time, they all follow their mother as one, wherever she goes.

Artificially incubated chicks that have never heard or seen a mother hen will even follow moving boxes or a flickering light. We can see from this that particular behaviours are instinctively pre-determined. A particular key stimulus triggers a particular reaction. It is also interesting that this readiness to follow only persists until about the eighth day of life. Chicks that have not become imprinted on a hen are no longer prepared to follow or to be led by a broody hen introduced to them after this time.

Fostering

Within the first week, as far as the mutual acceptance by the small chicks and a hen is concerned, the situation is still highly malleable. Although the chicks have imprinted very precisely on the voice and appearance of their mother in the first hours and days, it is always possible – when the circumstances dictate – to accustom chicks during the first week to another hen and to persuade a broody hen to adopt another's chicks.

For all the parties involved in these transactions, it is most advantageous to bring them together in the evening at dusk. In this way, they have the whole night to get to know, and become accustomed to, one another.

The Mother's Role

During their first excursions into their surroundings, the chicks do not leave their mother's side. As soon as one of them loses contact with the lively crew, it begins to cheep pitifully.

Good to know

The imprinting phase creates a very close relationship between the hen and the chicks, since the chicks' first imprinting influences normally relate to the hen and their siblings.

Good to know

It is important for the chicks that the new mother looks as similar as possible to the old one, which means, above all, having the same coloration. A good broody hen, however, will accept chicks of any colour, whether incubated by another hen or by an incubating device.

Good to know

The young do not have to learn this warning call, just as the hen does not need to learn the flight pattern of a bird of prey. Both the reaction of the hen to the appearance of a predator in the sky and the fleeing or freezing of the chicks in response to the hen's warning call are instinctive patterns of behaviour.

Finding food also has to be learned. Although the pecking impulse of the chicks is innate, the hen helps them by example to distinguish between edible and inedible objects.

This penetrating sound, the "cry of abandonment", is something we will often hear since, at a distance of more than about 10 metres, the chicks can no longer recognise their mother. The hen will respond immediately with loud clucking and lead the lost infant back to the flock.

Only when the mother hen has her little ones under strict control can she effectively counteract the dangers that lurk everywhere. A guiding mother hen is exceedingly protective and is always ready to fight to the bitter end. She faces up to any supposed attacker with furious determination. With head held high, feathers puffed up and a threatening look, she is such an imposing sight that even cats and dogs treat her with respect. And she is particularly aware of dangers coming from the air. If she sights the typical flight pattern of a bird of prey, she emits a penetrating warning cry which causes the chicks to seek protection near her or in a suitable hiding place or, if they are currently amusing themselves further from her, to freeze on the spot.

Once the danger has gone, an audible "cluck-cluck-cluck" releases the little feather balls to move again or draws them back out of their hiding places.

The reaction of the chicks to these two auditory stimuli can also be tested in chicks incubated artificially without a hen. All that is needed is to imitate the warning cry of the hen, and immediately, as if struck by lightning, the chicks press themselves down into the bedding litter and remain motionless until they hear the all-clear signal, the drawn-out, calm "cluck – cluck – cluck".

Learning

Chickens do not live purely by instinct. Like all higher life forms, they are extremely capable of learning. This ability is very important for satisfying a fundamental need, namely feeding. The young chickens have a certain instinctive drive to peck; that is, a natural interest in grasping small moving animals like spiders, flies, ants and objects the size of a wheat grain in their beaks and investigating them inquisitively. The fact that these small living and dead objects can be eaten as food is first learned from their mother. In particular the ability to distinguish between edible and inedible items is what they learn from the example provided by their mother, who leads them to particular delicacies and encourages them by picking items up, holding and letting them drop, to pick up and swallow such objects. This teaching is reinforced – and how could it be otherwise? – by music; that is, the "food calls" of the older birds.

Separation

Once the little brood have learned their lessons and have grown to the stage where they no longer need the constant protection of the hen, at about eight weeks of age, the family group breaks up, in some cases gradually and almost unnoticeably, but in other cases, from one day to the next. And then the otherwise nurturing mother hen suddenly drives her chicks away by pecking at them. Now, a new phase begins in the lives of our young chickens.

Closely guarded, these two hens are enjoying a sandbath. It serves not only for their sense of wellbeing, but also performs an important hygiene-related function.

Bodily Hygiene

From time to time, in the pauses between seeking food, egg-laying and other activities, chickens attend to the care of their plumage. With great skill, individual feathers are grasped in the pointed beak and put into order. The skin between the feathers is scratched and checked over, with noisy nibbling, for parasites.

A very special type of grooming is the *sandbath* or *dustbath*, which the animals usually perform in the early afternoon hours. To do this, they squat down in the sand, in dry, somewhat dusty soil or in the loose bedding litter. The dust and sand is kicked up with powerful scratching movements, remains lying on the animals' backs and then percolates slowly through the opened wings. Blissfully, they turn onto their side, stretch out wings and legs and burrow themselves ever deeper in. Finally, the sandbath is ended with powerful shaking of the body. Apart from enhancing the animals' wellbeing, it serves above all for removing annoying skin parasites, which, as the grooming process is completed, are shaken off together with the dust. For this reason, we should keep this possibility available for our chickens during the winter as well.

We cover this in greater detail in the chapter on housing. It is worth taking some trouble over this since a healthy plumage free from pests results in improved wellbeing.

Feed Preferences

In the previous chapter, we saw that chickens do not have a highly developed sense of taste, but rather a fine sense of touch for food. Chickens' preference for particular food types is determined less by their taste properties than by their structure, size, shape, hardness and surface texture. They prefer feed items that they can pick up and consume easily, whereas objects that are difficult and time-consuming to break up are not at all to their liking. Chickens live by a "peck and go" principle. For this reason, they like whole grains more than finely ground feed mixtures, since with the latter, they need much more time to fill up. Wheat is at the top of the popularity scale, followed by maize, barley, rye and, finally, oats. It is also noteworthy that in the case of maize, the animals must become used to the significantly larger grains before that cereal is included, just after wheat, on their scale of likes. As far as green fodder is concerned, thick-leaved, smooth-textured, tender plants such as rape, dandelion and clover, and cabbage leaves, juicy turnips and beetroot are liked best of all. But they reject tough or hairy parts of plants, such as cucumber leaves or fully grown grasses.

Good to know

Since envy over food is particularly strong in chickens, we should ensure we provide sufficiently large feeding containers, as otherwise the lowest-ranking animals will often be left out and will do less well than they could.

Fighting

At the age of about two weeks, the chicks, which have now become quite boisterous, are already engaging in early trials of strength. Beating their wings, they run and jump at one another, wait furtively for brief moments and then, at the next instant, turn away abruptly and flap towards the next playmate. And only a few days later, it can come to pass that one of these little fighters deliberately pecks his opponent right in the face and the other individual withdraws looking disconcerted. These are the first beginnings of early social status disputes.

But things only turn serious when sexual maturity sets in, after which each bird must fight it out for its position within the flock. Such pecking-order fights are often harsh and merciless. Not infrequently, they end with bloodshed, though they are never deadly. After a few days of unrest and fuss in the flock, it is all over with. The winners and losers have been decided. From this point on, each animal knows its fixed place in the hierarchy of the flock, so that only in exceptional cases do any further status fights need to take place.

Hen Fights

Simple disagreements between the hens are cleared up just with short, powerful pecks of the beak, whereupon the lower-ranking individual soon gives up and seeks safety by fleeing. If, however, it turns into a real fight, which most often arises between two animals of very close rank, the two antagonists approach each other initially with low drawn-out threat sounds. Then they take up their positions close to each other and with an upright posture and neck feathers ruffled up. Suddenly, the attack takes place as they jump up and peck at one another. The aim of these

Good to know

In the case of cock fights, we can distinguish between the usually harmless feigned fighting, rather like a medieval joust, and the hard, often bloody fight for dominance in the flock.

beak blows is mainly the back of the opponent's head, but the face and the comb are not spared either. In most cases, the fight ends just as suddenly as it began. The defeated hen cowers and, with this posture of humility, shows its readiness to submit.

Cock Fights

A fight between cocks has a far more elegant appearance and the sequence is very different from that of hens.

During a feigned fight, the birds circle round each other, initially playfully, but with a threatening demeanour, and crow their arrogant challenge to each other. Then, feigning courage and determination, they take a few steps toward one another, only to turn away again at the last moment. If the pair accidentally or deliberately get too close to each other, the elegant, playful skirmish becomes unavoidably serious.

At the same time, or in alternation, the waves of attack roll in from both sides with the opponents dealing each other hefty wing-blows and jumping up at one another. In the process, they attempt with heavy kicks to hit their opponent in the chest or head with their sturdy, sharp spurs. The defender tries to repel the attack of the other with a crouching posture and with his wings spread out like a shield. Between the assaults, the two cocks pause for breath with head held low and neck stretched out, standing opposite one another in ambush posture. As exhaustion sets in, they even attempt, rather like a severely battered boxer, to find sanctuary briefly under the wings or between the legs of the other animal.

The fight is over when the weaker of the two finally turns to flee with neck feathers splayed out.

Fear

Immediately after hatching, chicks are completely unafraid. But after a few days, we can see the first signs of fear and fright in them. Unfamiliar sounds and hasty movements, in particular, make them fearful. This can lead to dangerous situations for the young chicks because if they press themselves together, huddled in fright in a corner, there is a risk that they could suffocate each other. To prevent this danger, we use "chick rings" during the rearing phase and no right-angled boundary walls.

Adult animals are also sensitive to and mistrustful of all strangers and the unfamiliar. For example, our ten-member chicken flock – otherwise used to our excitable dog, cats from the neighbourhood and our rabbits – once became very agitated as our small son appeared in the henhouse with a furry animal toy under his arm. This new phenomenon was not only unfamiliar to them, but clearly also triggered an instinctive fear of a hereditary enemy, the fox. In a similar way to the flight pattern of a bird of prey, mentioned above, chickens have an in-bred fear of everything that resembles this predator. From this we can see that, due to their poorly developed cerebrum and their consequently restricted learning ability, chickens still react very strongly to inherited key stimuli.

Bad Habits and their Remedies

There is no doubt that bad examples can easily be copied and become established habits and that these need to be nipped in the bud. This rule applies, to a particular extent, among domestic chickens.

If we ever notice one of our egg-laying hens successfully interfering with the plumage of a neighbour and plucking out her

Feather-pecking is one of the main problems in keeping chickens. The bare patches on the neck, back and cloaca region are clearly apparent.

Good to know

Unknown sounds, objects and living beings can put them into a real panic. They are especially afraid of moving objects with a furry texture.

feathers, or showing a taste for her own eggs, we need to be extremely vigilant. Due to the strong imitation drive among our birds, it will not take long until we have not just one egg-eater or feather-pecker in the flock, but many. Apart from recognising these bad habits and taking suitable action to prevent their spread, it is far more important to know the causes and some preventative measures in order to root out such curses.

Feather-pecking

Feather-pecking or *feather-eating* is one of the main problems of this type, and not only in commercial chicken keeping. The final stage of this vice is usually serious injuries to the harassed animals. And as a further consequence, it can even lead to cannibalism.

Typically, it begins harmlessly. Out of curiosity, boredom or for some other reason, a hen approaches another and cautiously plucks at a protruding feather. It is well known that a particular stimulus for this behaviour is provided by slightly protruding tail or neck feathers, which are often worked at by the culprit for minutes at a time. Since, for unexplained reasons, the pecked animals only rarely defend themselves or take flight, this stereotypical play turns into more intensive pecking.

Finally, the activity is not directed only to the tender feather tips, but also the attachment site, so that eventually the feather works loose. Now, the animal with the pecking habit really starts to enjoy its behaviour. The attacks on the victim become ever more targeted, so that soon, bald patches appear in its plumage. The favourite regions for these attacks are the neck, the back and the region of the cloaca. But it doesn't end there. Due to the violent tearing out of the feathers, eventually bleeding wounds are formed which act like a red rag for all the others. Soon the victim has more than just the feather-eater to deal with and now has a bloodthirsty flock after her. A cannibalistic pursuit of the poor creature begins. And it is not unknown for the doomed animal to have its innards dragged out of its cloaca while still alive.

Causes

The initially harmless-seeming feather-pecking may have genetic origins or be triggered by housing or feeding faults.

Feeding faults may lie in unbalanced feed rations. In particular, a lack of minerals and the absence of calcium and sodium are thought to initiate the fateful behaviour. There is also scientifically backed knowledge about suitable or unsuitable forms of feed. Pelleted feed, for instance, due to its coarser structure and the rapid consumption rate, results in shorter feeding times. On bad weather days and in the winter when we can't let the animals go out, they suffer from boredom and irritation, so that the risk of feather pecking is greatly increased.

The main causes mentioned above also carry within them the answers for suitable countermeasures.

Prevention and Corrective Measures

If we give our animals a stimulating, multi-ingredient and variable main meal, enough greenery with it and possibly also additional minerals, then we will have achieved a great deal towards nipping the occurrence of this creeping and pervasive vice in the bud. We also give our birds small quantities of cereal grains separately in the litter, so that they can dispel their boredom by busily scratching about.

The second cause can be dealt with by avoiding over-crowding of the henhouse. Based on the principle that quality is better than quantity, we should house one or two fewer hens than the norm suggests. Added to this, there should be a sufficiently large run and – particularly for the winter and periods of poor weather – an adequately sized scratching area with deep, clean litter. These matters will be dealt with in detail in the chapter on housing.

The starting point for appropriate and rapid countermeasures is constant monitoring of the individual animals for signs of changed behaviour patterns. A feather-pecker caught in the act is separated immediately from the flock and an attempt is made, with a suitable diet containing especially large amounts of minerals, to free her from her bad habit. At the same time, we should always consider whether our feeding and housing system fully meets the requirements mentioned above. And, finally, unimproved repeat offenders are destined for the table.

But we should never allow ourselves to be tempted to give the often badly dishevelled-looking victims over for slaughter because they have become somewhat unsightly. If we did, eventually only the feather-pecker would survive. First of all, though, we should treat the wounds inflicted on the abused hens with disinfectant.

Some retail outlets also have special oil-containing preparations which, when applied to wounds, prevent further bullying of the injured animals by others. The effect of these preparations is often only short lasting and daily re-application over several weeks is arduous for the chicken keeper.

Another workable countermeasure that has been hailed by some experts is "debeaking", in

Good to know

Among the cardinal errors leading to feather-pecking are overfilled huts, a shortage of run space and insufficient scratching area. The absence of a cock also has an unfavourable effect because he exerts a decisive influence on the social harmony of a flock.

Good to know

To protect them from further attacks, the injured animals are separated from the remaining flock until the wounds have healed tolerably and the bleeding skin areas do not entice the otherwise well-behaved hens to peck at them.

Good to know

Healthy human common sense should itself forbid us from feeding egg shells as a replacement for shell limestone. Surely, offering egg shells, even broken up into small pieces is a cause for this – put scientifically – oophagous appetite among hens.

which the upper and lower beak are shortened with a glowing-hot knife. This should only be done where indicated on veterinary grounds. Apart from the questionability of this method with regard to animal welfare, the severely shortened beak no longer allows proper preening and significantly impairs the animals' sense of touch. The latter might be unimportant for hens kept in intensive conditions with ready-made feed mixtures in the form of meal, but for our run-housed hens, a reduction in their sense of feel is a serious handicap.

The temporary application of plastic blinkers on the upper beak of the hen seems to be an alternative worth considering. Our animals, we have found, are not so keen on the idea.

With our free-range keeping arrangements, we have hardly had to resort to this unusual measure.

Egg-eating

The next main vice, egg-eating by the hens, is a no less distressing experience according to some chicken keepers. If the keeper discovers, one day, that the egg yield is declining although the hens must be laying eggs, as revealed by daily examination of their bellies, he or she will soon find out, after close observation, that some of the birds are tampering with their own produce. In this case, also, rapid action is needed, since the hens' drive to imitate each other is great. It can even develop to the stage where a hen, excitedly expectant of the delicacy that is about to arrive, patrols impatiently in front of the nest of a hen who is in the process of laying, in order to feast on the still warm egg as soon as it arrives.

During a study visit in Scotland at a poultry research institute, we witnessed how almost the entire flock of hens gathered round – indeed, veritably mobbed – a fellow hen who was ready to lay, minutes before the egg appeared, in order to throw themselves at the next moment upon the object of their appetite.

Causes

The causes for this vice have not yet been clearly identified. Suspicion rests on the negative influence of calcium deficiency and the associated laying of thin-shelled or even shell-less eggs.

Mislaid eggs – that is, eggs that are not laid in the intended nests – seem also to provide a certain stimulus.

And, as with feather-pecking, a dull day lacking in variety also encourages this habit. Finally, the age of the animals appears to play a role in the occurrence of both of the vices of feather-pecking and egg-eating. With increasing age, hens become more quarrelsome and more aggressive toward their fellows and their surroundings.

Countermeasures

Practical experience and, particularly, older text books offer numerous, but mostly unsuccessful, recipes for dealing with the egg-eating habit:

The hen should receive no food during the day, but be given some porcelain eggs,

> *which she will attempt to feed on. In the evening, the hen should then be fed and the same procedure repeated for several days until she leaves the artificial eggs and, finally, natural eggs alone.*

Or,

> *Place really hot hard-boiled eggs in front of the egg eaters so that, when they peck at them, they scald their beaks and then, in future, will feel similar wariness at the sight of eggs.*

The simplest method is to collect the eggs more frequently and to put more bedding into the nests in order to avoid broken eggs, since, as has been shown, this habit often begins with the accidental acquaintance of a hen with eggs that are dented or broken open. The provision of roll-out nests can also be helpful, since with these, the hens can no longer reach the eggs after laying. When we can find no remedy, slaughter is the last solution and, for the chicken keeper, certainly not the worst.

Egg Mislaying

Egg mislaying is one of the most harmless but annoying bad habits. For one thing, it is very time-consuming to find the eggs, which have often been hidden very artfully in the hut or the run, and for another, they are often dirty or damaged.

Good to know

One should also make sure that there are enough nests so that the lower-ranking hens are not compelled to search for a secluded place outside.

Prevention and Corrective Measures

There are some preventative measures we can take. For the location of the nest boxes, we should not choose the sunniest part of the hut, because the hens tend to withdraw into a shaded environment when they are ready to lay. We need also to ensure that the nests are clean and free from pests, so that the animals can carry out the daily act of delivery without being molested.

We can help young hens who are just entering into their laying stage to find the nests by placing porcelain, plaster or white wooden eggs in the nests, since hens are well known to prefer nests that have previously been used.

Somewhat more trouble, but also successful, is to feel the young animals and to include those you suspect of being ready to lay in the nest. Soon afterwards, in need of relief, they will seek it out themselves.

However, if we have a hen in the flock who does not wish to be persuaded and is particularly successful at hiding her eggs, we should accept it with good humour. After all, it is not Easter every day. We should never let our annoyance lead us to use the following remedy, which was once quite common:

> *The hen is kept in the henhouse until she shows the willingness to lay, whereupon a few grains of salt should be placed in her oviduct. Driven by the stimulus of the salt, she hurries to her concealed nest and reveals her hidden eggs.*

CHAPTER 5

Keeping Laying Hens

Once the decision has been made to keep chickens, it is important to ask oneself about one's motives. Do I want chickens purely for the enjoyment of keeping poultry, for fresh eggs or for the tender, tasty meat, or perhaps for all three reasons? In general, the latter will be the case and here the chicken keeper will find himself facing a hard test, namely when the time comes to slaughter one of his trusted egg providers. A second important question that must also be considered is how much space one has available and then, thirdly, how great the maintenance demands of one's charges will be – i.e. how much time one must, and is willing to, invest in them.

Which breed of chickens to keep, how large the flock should be and what size the henhouse and the run must be made depend on these three factors. In the following section, we assume we are dealing with a chicken keeper who wishes to combine the pleasurable and functional aspects – that is, the joy of owning chickens, together with the enjoyment of the meat and the eggs – and who also has sufficient suitable space for poultry keeping. We also assume that no existing structures can be used as a home for the animals and that no objections from neighbours are to be expected. The possible use of old buildings should essentially be based on the recommendations made here. These conditions being met, we can begin planning for about a dozen birds, 11 or 12 hens and one cock.

The Henhouse

If we look around in the countryside to see how domestic chickens are usually accommodated, we can find many possible housing types, from draughty corrugated iron huts to the most luxurious of dwellings. But what we want is the most suitable accommodation possible at an appropriate price where the animals will feel at home and the daily maintenance work is as easy as possible. The henhouse is, after all, the place where the major part of chicken keeping takes place.

Building Materials

Wood, masonry or a combination of both can be used as materials for building the henhouse. Modern building materials allow for simple construction at a high-quality standard. From the standpoint of warmth, both types of material offer satisfactory solutions. In the end, the decision is a matter of taste and which materials one feels most comfortable using.

Hut Size

The dimensions of the hut are, of course, dependent on the size and breed of the flock to be kept. Heavy animals naturally need more space than light types and full-sized hens more than miniatures. This might well sound obvious, but it is often overlooked during planning and in practical chicken keeping. As a rule of thumb for a henhouse with a run, for a medium-weight breed, it is reckoned that there should be three hens per square metre of hut space, including the space required for perching rods and a droppings pit. It is possible to increase the number of hens, even without a run, but this is where we start to enter the realms of commercial poultry operations.

The floor area of the henhouse should be shaped as square as possible, to make the design as compact as it can be, which offers corresponding advantages from the energy standpoint. However, these rule-of-thumb figures reach the limits of realistic practicability with an extremely small flock of, say, just three animals. After all, a hut with a floor area of only 1 x 1 m can barely accommodate equipment, feed sacks and other items as well. For a sensible arrangement, it should be planned, with

Some facts and figures on henhouse design	
Hut size	1 m² per 3–4 chickens
Run size	10–20 m² per chicken
Laying nests	1 nest per 3–4 hens
Perch length	1 m per 4–5 animals
Feed trough length	10–15 cm per chicken
Feed requirement	115–130 g per bird/ day
Water requirement	200–250 g per bird/ day

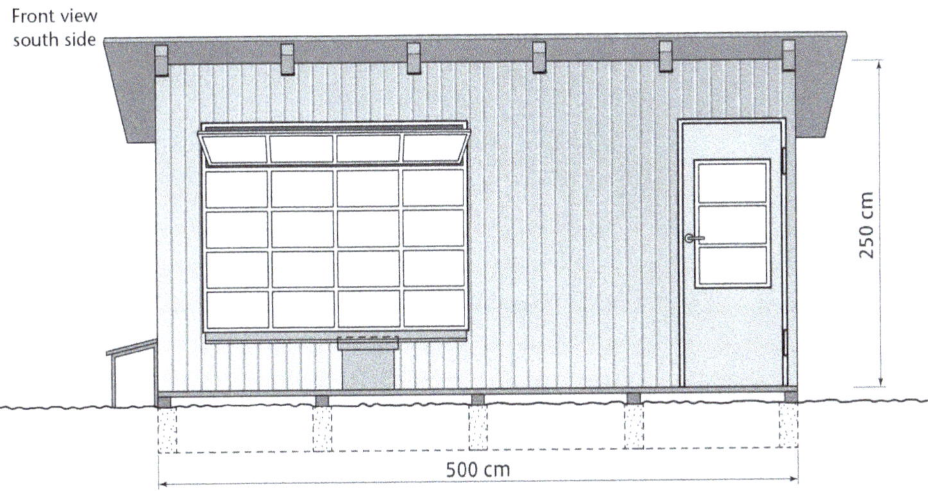

Front view of a henhouse: clearly shown is the large window, which will provide sufficient light and air.

a hut size of up to 4 m^2, to provide a separate add-on section for equipment and feed storage.

For our own twelve hens and a cock, we need an overall area of approximately 4 m^2 as a minimum. We should also provide the aforementioned space for equipment and feed as an add-on or a section of the same hut. In our suggested layout, three laying nests are also included. The overall planning should always start from the realisation that it is no error to design our henhouse to be somewhat more spacious than the rules of thumb assume. And we have indeed consistently applied this maxim in our planning recommendations.

From the construction standpoint, we have approached the task like the building of a family house without a cellar. What is needed, then, are a suitable site, foundation strips with a floor slab, walls, doors, windows, a roof, building material and tools.

Location

A suitable location depends initially on the local conditions, but it should be as flat as possible and close to the house. The window should face toward the south or southeast so that as much daylight as possible can enter and so that the hut does not overheat in high summer, as it easily could if the windows were placed in a southwest-facing front. A location that is as well protected from wind as possible is highly advantageous, since hens are very sensitive to wind and draughts. The site for the building and the run should be dry, since otherwise a constantly damp henhouse can be expected. There is also always a fundamental choice to be made between fixed and movable henhouses.

Movable henhouses can naturally only provide housing for a small flock, unless a large house can be obtained by converting a lorry trailer, builder's site hut, a shepherd's hut or something similar. Housing of this kind would

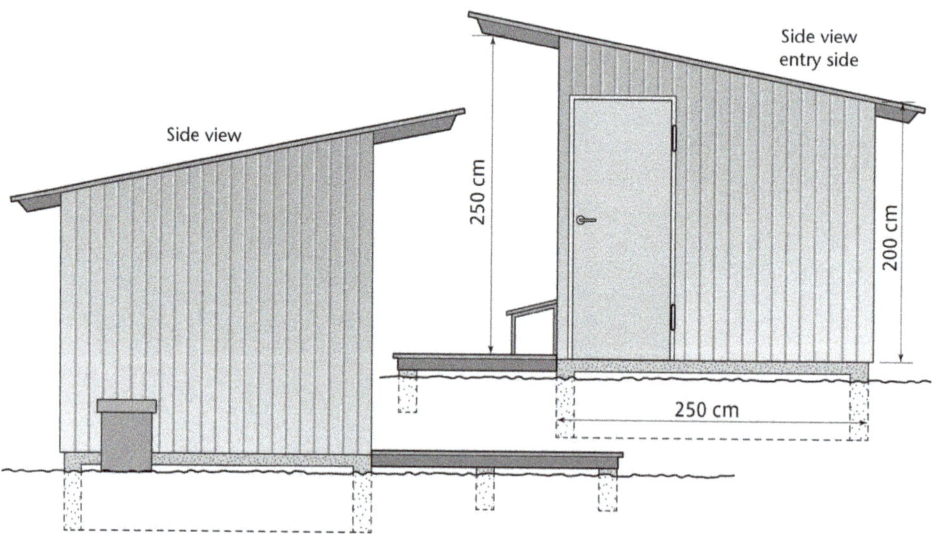

The henhouse in an exterior side view.

only be useful for shepherded chicken keeping without any fencing, which is something that not many people would find enough time for. If it is intended to construct a permanent building, then of course, above a certain size, planning permission might be needed.

In any event, information about the legal conditions and the paperwork to be submitted can be obtained from the relevant local authority. In our conceptualisation and planning of a permanent structure as set out below, we assume a generously dimensioned building with a floor area of 5 x 2.5 m, and the rule-of-thumb values given are to be taken as minimum sizes.

Good to know

The concrete floor slab offers the advantage of simple cleaning and disinfection of the hut, although it does not permit any ventilation from under the floor.

Henhouse Construction

Foundation

For the foundation or, better still, the foundation strips, a foundation trench approximately 30 to 50 cm deep and 20 cm wide should be dug out and this should be boxed out to about 30 cm above the natural ground surface. Into this trench, we should first place a gravel filling, to about 10 cm depth, and then fill it to the upper edge of the box with concrete. The inclusion of thick wire and a variety of metal rods, if we can get them, is useful to increase the strength and is therefore to be recommended. If it is planned to construct the walls of wood, then suitable anchor bolts should also be included for the joists that will be mounted on the foundation.

The floor of the hut can be built in a variety of ways. What is important, in any case, is that the floor level should be about 20 to 30 cm above the ground level, so that the hut stays dry. In general, a gravel filling is laid first, onto which either concrete is poured as a closed slab or fine sand is laid, followed, for example, by bricks. In both cases, the floor should be provided with a slight incline and suitable means for water to run away. If, rather than a closed concrete slab, bricks, which can also be pointed with lime mortar, are laid on a bed of sand, then the hut floor will be kept dry by a continuous updraught from below. This also promotes the biological decomposition of the droppings in the bedding, thereby improving the overall conditions in the henhouse. A precondition for this is the inclusion of ventilation pipes at the level of the gravel layer, perpendicular to the foundation strips, enabling the desired gas exchange through the gravel layer.

Walls

For the building of the hut walls, either brickwork or a wooden construction suggest themselves as alternatives. Masonry walls have the advantage of good heat-storing capability and, if suitable materials are used, do not need any heat insulation. However, this building method is difficult for DIY enthusiasts without experience and is also usually more expensive. Wooden constructions are ideal for the DIY builder and if the hut is built in a lightweight construction of this kind, the foundation can be made correspondingly less robust, which also saves costs.

The wooden construction is set on the foundation, which projects above the floor, and is firmly connected to it by means of the lower joists with the embedded anchor bolts.

The wooden beams or the timber framework is faced on both sides with wooden boards and the intermediate space is filled with insulating material.

For tho outor wall, according to taste and the structural conditions, weatherboard or board-and-batten cladding can be used. In each case, it must be ensured that rain water can readily run away and cannot form water pockets which would cause the wood to rot. The inner facing should be created, where possible, with planed tongue-and-groove boards or water-resistant chipboard or fibreboard to simplify cleaning. For the protective treatment of the wooden facing in the interior, toxic preparations need to be avoided, since residues could find their way into the eggs. As alternatives, non-toxic and environmentally sound chemical products or biological substances such as linseed oil preparations could be used. One could opt to dispense with a treatment, although this may have the consequence that parts of the wooden facing would need to be replaced sooner than otherwise.

In the case of masonry, coating in the spring and autumn with whitewash, which is a proven measure against pests, would suffice.

However, after a few years, this interior coating should be removed rather than over-coated, since, above a certain thickness, it prevents the important gas and moisture exchange through the wall. The finish of the outer wall is a question of taste alone. Coating with whitewash seems to be the simplest and most economical method here also. A lime cement exterior render looks smarter, but requires a higher degree of skill to apply and also costs rather more. One thing that should be avoided in any event is something that is so often used in housebuilding today, namely installing a damp-proofing layer into the wall in the form of a film. This almost completely prevents the exchange of gas and moisture. The result would again be a constantly damp henhouse with the consequence of a poor climate inside.

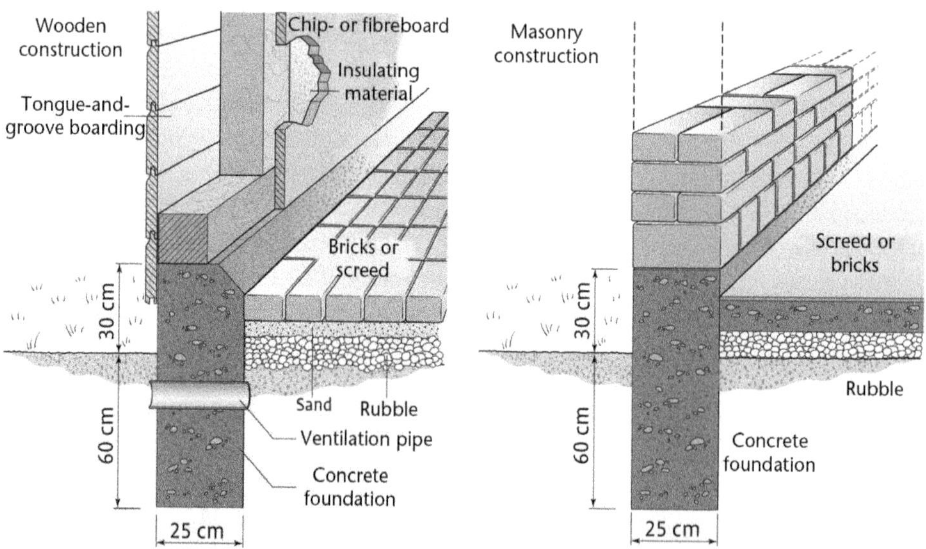

Shown here are two alternatives for the foundation, the floor and the wall constructions.

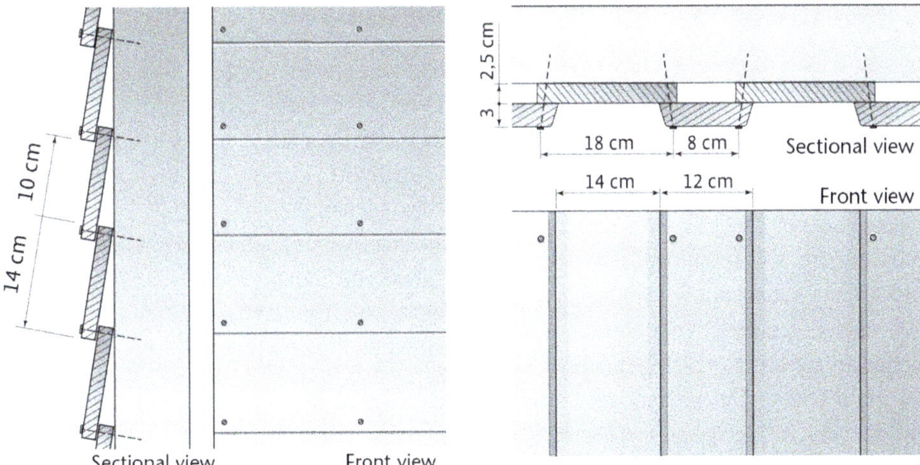

On the left, weatherboard cladding, on the right, board-and-batten cladding for the outer wall of the hut. Both are suitable and the choice is purely a matter of taste.

Roof

Suitable roof shapes for a henhouse are the single-pitched roof or the double-pitched roof. Both forms and also certain intermediate forms offer satisfactory construction solutions.

The simplest and most economical variant is the single-pitched roof, although the different wall heights must be taken into account from the start. The most straightforward possibility is simply to place the roof joists on the side walls of different height at suitable spacings, to nail the roof boards onto these and to seal the boards, with tarred roofing felt as the roof covering.

The disadvantage of this method is that the roofing felt can easily be torn or holed when being applied, so that moisture penetrates at these points and the wood can start to rot. In addition, huts with this kind of roofing become very hot in summer since no air circulation is possible between the roof edging and the covering. This is, incidentally, a further reason why the wood can rot from underneath, due to the high air moisture from the hut interior.

It is better, although more expensive, to construct an insulated roof which is closed off from the outside with a skin of fibre cement boards. Roof tiles are not so good for single-pitched roofs since this roof type normally has only a slight inclination. They are also more expensive.

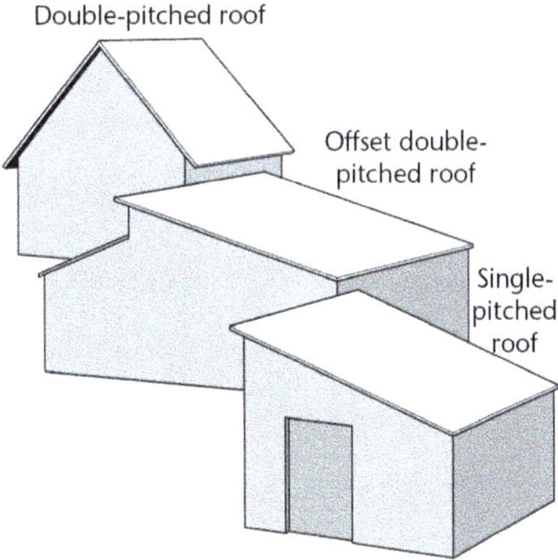

All three roof forms are essentially suitable for a henhouse.

For many people, the double-pitched roof is surely the more appealing design, since it usually harmonises better with the nearby house. Depending on the roof angle chosen, there are two variants in this respect: with a relatively shallow roof angle, an insulated roof with fibre cement boards or tiles, or, with a steeper angle (above about 30°), a non-insulated roof and the inclusion of a suspended ceiling.

Insulation against cold and heat can be achieved by storing bedding litter – such as straw bales – on the suspended ceiling. In this way, it is possible to kill three birds with one stone, as it were. One achieves an attractive construction, good insulation and a suitable storage space. The eaves should be dimensioned so that they correspond to the proportions of the henhouse and also protect the walls against rain hitting them, while not preventing sunlight from entering.

Henhouse Climate

Ventilation and lighting are particularly important aspects of henhouse planning.

Ventilation

Hens benefit from plentiful fresh air, since they need significantly more oxygen than other livestock. This would not be a problem if they were not, at the same time, exceptionally sensitive to draughts.

The solutions to these issues become more difficult and complex as the size of the flock increases. This applies to the same extent for outgoing air with which the used air, the excess moisture and henhouse germs are carried away. With larger flocks (over 100 hens), built-in extractor ducts are used which draw up the used air close to the floor and convey it via the roof to the outside.

For smaller numbers of birds, a variety of air inlet and outlet flaps with no mechanical contrivances such as fans, are entirely sufficient. Use is also made of the physical phenomenon that cold air falls to the floor and warmed air rises. Following this principle, in the case of a single-pitched roof, flaps or sliders for ventilation are mounted at the rear side – i.e. the lower wall. Since, however, the droppings pit is usually arranged at the rear side of the

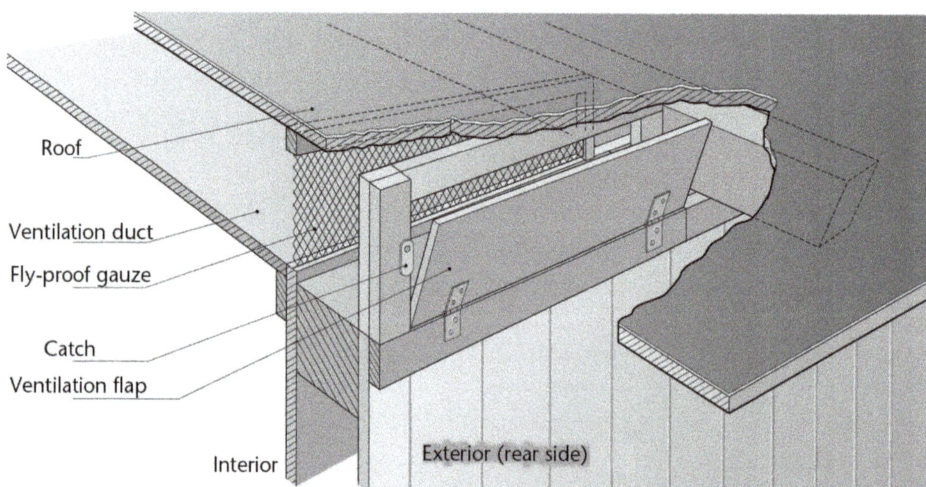

Ventilation flaps of this type provide the necessary fresh air supply.

hut with the perching rods placed over it, the animals would then be constantly sitting in a draught due to the downward-flowing air. We can help this situation by covering the roof spars from underneath over the floor area of the droppings pit with wood so that the cold air initially flows in through this channel and only flows down in front of the droppings pit. The sinking fresh air now becomes distributed in the hut before reaching the birds.

The used air and the unused part of the cold air is warmed in the lower regions near the floor and only then tends to escape upwards.

Suitable outlet flaps for this exhaust air should be provided in the front of the hut under the roof or in the upper part of the window.

A similar arrangement can be incorporated with the double-pitched roof design. In this case, the ventilation inlet should be included in the upper part of the windows or with suitable arrangements at the height of the eaves and the air outlet should be arranged in the roof ridge.

Two rules of thumb should be applied to ventilation:

- No ventilation device should lack protection against feed or livestock thieves. Sparrows, mice, rats and weasels are best kept away with tight-meshed wire netting of sufficient strength.
- It is better to have numerous small ventilation units than a large opening window. This affords the chicken keeper the possibility of optimally adapting the ventilation to the stock numbers and the weather.

Light

Lighting is the second important component in the design of the environment in the hut. For this purpose, if possible, a window should be provided in the southeast side of the henhouse, its size being dimensioned so that dark corners are formed in the hut. This principle is important since sunlight, which kills bacteria, cannot penetrate into such corners and it is also to be expected that the hens will prefer to lay their eggs there. As a general rule, the window area should be approximately ⅓ of the floor area.

Of importance apart from the size of the window area are the ground plan of the floor and the orientation of the hut. Square floor areas up to a certain size need smaller window areas than rectangular layouts, provided the height of the window is sufficient to illuminate the rear part of the hut adequately. With a wall height of 2.4 m, this limit is reached with a hut depth of about 4 m. So that the front part of the hut also receives enough direct sunlight, the window sill should be at about 40 cm height so that a window that is 1.6 m high has its upper edge at a height of 2 m.

Normal window glass is sufficient, although with this, shutters or something similar should be provided as heat-loss protection for the colder times of year. More comfortable, of course, but also more expensive are insulating glass panes for additional heat insulation. A third possibility is to install an additional glass pane by way of double glazing for the winter.

Good to know

Sunlight has the great advantage compared with artificial light that it costs nothing and is particularly beneficial to the wellbeing of the animals.

Good to know

It is, in any case, advisable to place a wire frame over the whole window to prevent the chickens from leaving droppings on the sill or escaping through the open tilting window.

For relatively small henhouses without any sophisticated ventilation system, the windows should be constructed so that they can be opened.

A useful solution is a tilting window in the upper third, the triangular opening angle being filled in on both sides with walls of wood or concrete which prevent the sideways penetration of draughts. It would be ideal if the whole tilting window could be completely removed during the warm summer months.

Door

The door is often neglected during planning. To save oneself problems later, it is important to observe the following points. Firstly, the door should be absolutely leakproof – perhaps made of strong tongue-and-groove boards – and perhaps containing additional insulation to match the hut walls. It should close tightly and, for that reason should be built, as far as possible, twist-proof and be hung in a solid frame, opening outwardly. Lastly, one should not forget that, when clearing out, the happy chicken keeper who has to balance each forkload of droppings from the back part of the hut all the way to the door, because the door is not wide enough for his wheelbarrow, quickly starts to feel less cheerful.

Equipment Room

We regard it as exceptionally useful that a suitable equipment room is provided in the plans for the henhouse as a place to store equipment, feed and bedding litter. All the important things that we need to have quickly to hand have their place here. Since it cannot always be assumed that all the chickens are hand-tame, even though in many cases we are forced to handle the animals, a catching hook or a collapsible catching gate is indispensable. For cleaning and for spreading the bedding, the minimum equipment is a dung fork, a shovel, a broom, a stiff-bristled brush and a wheelbarrow. Also needed are a basket for collecting eggs, a knife for chopping up garden waste and a hammer, pliers and a screwdriver, as well as screws and nails, for small repairs.

Part of the bedding litter can also be stored in the equipment room. This has the disadvantage, however, that pests and particularly various rodents, which help themselves from the feed bags, can quickly colonise this area.

The minimum equipment should look something like this. This is based on the assumption that the chicken keeper also has available the normal range of household equipment.

If it is necessary, for space reasons, to store the litter in the equipment room, it is always only small quantities that should be stored so that they are quickly used up and so do not offer a long-term hiding place for feed-stealing creatures.

Good to know

If the henhouse does not have an electrical supply, a normal but powerful torch for checking the flock as dusk falls in the autumn and winter is an essential piece of equipment.

Hen Door

If we wish to allow our hens constant access to the run during the day, we need to build a hen door into the wall with a flap or a slider that can be closed from outside. The opening size should be approximately 30 cm in width and 40 cm in height, whilst for miniature hens, these can be about a third smaller. For larger flocks, correspondingly larger such doors should be provided. Again, it is important here that no draughts can arise. It therefore makes sense not to place the hen door opposite the window. A suitable point is under or close to a window. During cold weather, one could go one step further and, in front of the hen door, provide a windbreak box which can be opened on two sides, depending on the wind direction.

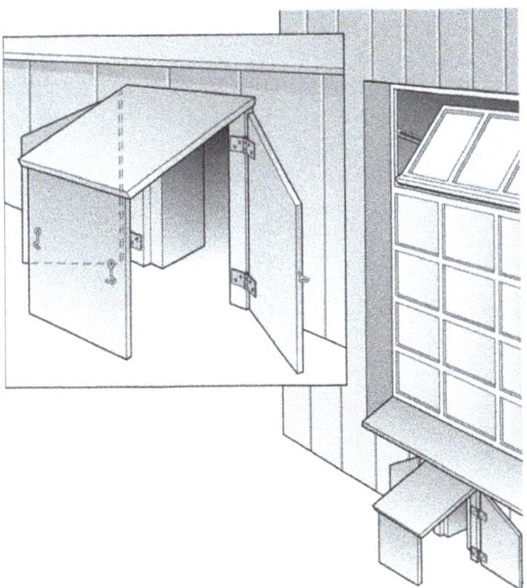

It is well worth including a windbreak box for the hen door because chickens are very sensitive to draughts.

So, summarising the points for planning the hut:

- Find a suitable location.
- Match the height and floor area to the flock size.
- Match the building materials to one's own DIY skills and/or budget.
- Ensure adequate insulation.
- Provide sufficient ventilation and illumination.
- Select a suitable roof shape.
- Design the access door to be wide enough for a wheelbarrow.
- Provide for a draught-free hen door.

Helpful tip

The henhouse should be laid out so that it is not necessary to cross the run to gain entry. This will save mess from dirty footwear.

Equipping the Henhouse

Once the hut itself has been completed, the interior fittings can be attended to. It makes good sense to equip the henhouse as nearly like the natural habitat of the chicken as possible. It should therefore be considered that a hen needs to move about and now and then will lay an egg – if possible, daily – for which purpose it needs a suitable nest. It also likes to have a sleeping place reserved, loves clean food and water, likes to scratch about on the ground and also to be able to "bathe" in dust.

The chicken keeper must provide for all these needs. And so that it does not become burdensome, one should equip the henhouse in a way that is as functional and labour-saving as possible. It has often proved to be the case, particularly where the feed and water containers are concerned, that those made by the specialists are preferable.

Perches

In the wild, chickens like to spend the night in a raised perching place, for example in a bush or a tree. This habit, or rather instinct, has not been abandoned by the domestic chicken, even though it is protected against foxes and other predators by the enclosing henhouse. For this reason, a well-equipped henhouse should not lack perching rods, which are usually positioned above the droppings pit. Since a relatively large amount of droppings collects under the perches in the droppings pit, it is advisable to protect the pit with a wire mesh so that the chickens cannot get into it, with the associated risk of diseases. Another possibility that is well suited to small flocks is to arrange droppings trays under the perches, with a slight incline toward the front, from which the droppings can be scraped at regular intervals so that here also, contamination of the hut and the associated risk of the spread of pathogens are reduced.

Droppings boards with a depth of more than 1.5 m from front to back are not suitable, however, since they are hard to clean in the rearmost part and the animals roosting on the rear perch are difficult to check.

The spacing of the rods from the wall and between each other should be 35 to 40 cm depending on the breed of chicken so that the animals' tail feathers do not become crushed and they are not able to molest one another constantly. For each bird, a perch width – also depending on breed – of about 20 to 25 cm should be allowed, which equates to about four or five hens per metre. Due to the greater ease of cleaning, the perches should lie loosely in suitable mounting timbers, so that, with a width of 4 to 5 cm and a depth of 3 cm, they should each be sunk by about 1.5 cm into an exactly matching depression. For hygiene reasons, it is usual to use smooth planed rods with a rectangular cross-section and with slightly rounded upper edges. It is also a good idea to oil them when new and after each cleaning, or to treat them with a

Just like its wild ancestors, the domestic chicken also likes to rest in a raised place.

non-toxic glaze, which keeps the surface smooth and makes it harder for mites to find a home on them.

Particular attention should be paid to the design of the perches and the droppings pit or droppings boards to ensure that all the parts are easily dismantled and removed, since this significantly facilitates cleaning and any repair work. This principle applies also for most other henhouse equipment items, which will be dealt with now.

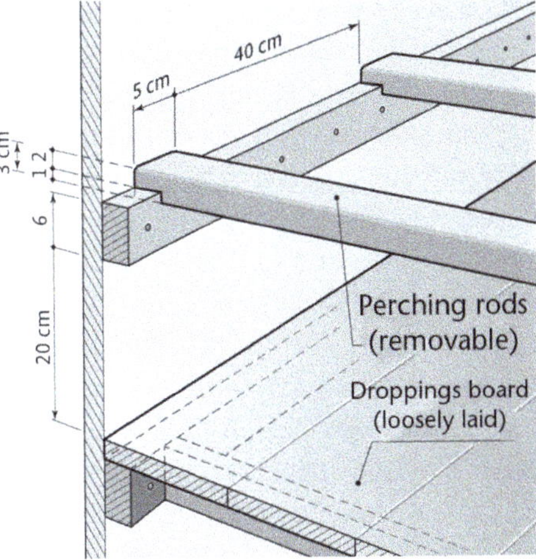

Perches and droppings boards should be easy to remove or dismantle, so as to simplify cleaning.

Sleeping Quarters

If we want to go a step further and give our birds a cosy place for the night – particularly in the colder months – we can close off both sides of the perching rods upward with boards or panels and provide a roller blind or a curtain on the front. This creates a little private space which, with its enclosed volume, can better conserve the chickens' own body heat. This design might seem a little excessive to some, but it has some real advantages, as mentioned, during the winter months. A construction with droppings boards can be adapted particularly easily to include these details.

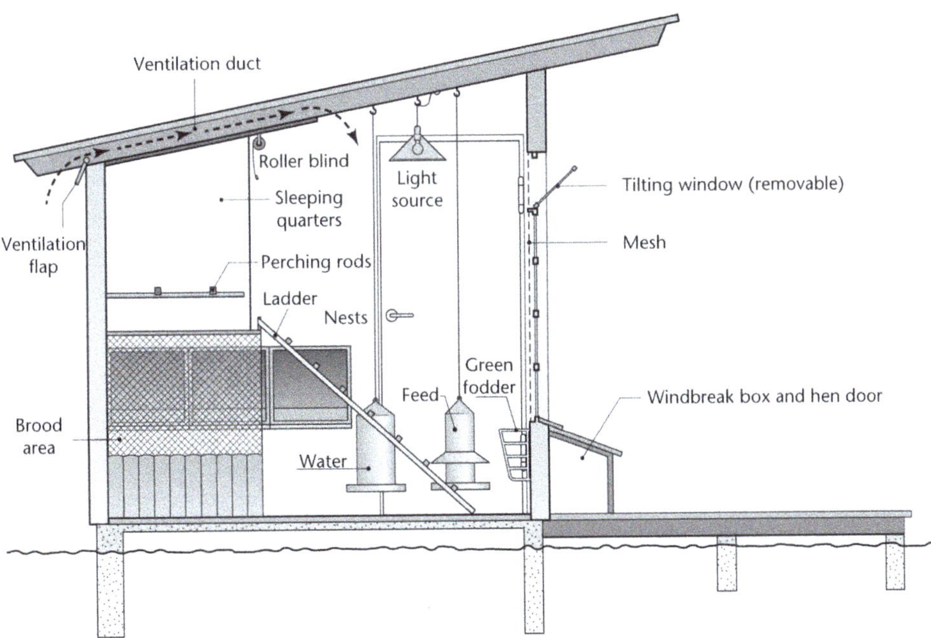

Our fully kitted-out henhouse seen in cross-section, showing, in particular, the ventilation principle.

Nests

In the preceding section on henhouse construction, we mentioned that the hut should be sufficiently bright over the whole width and depth since otherwise the hens lay their eggs in the dark corners. We should therefore provide dark corners for the hens precisely where we can remove their eggs easily and without damaging them. Basically, a used orange box fastened with its base to the wall of the hut and padded with soft material is sufficient for this purpose.

A fundamental distinction is made in nest design between *individual* and *family* nests. Human ingenuity has developed the individual nest to a highly technical standard; an example is the many forms of trap nest, although we will not concern ourselves with these in detail here. The basic principle is shown in the drawing. Trap nests have the advantage that one can find out, if necessary, which hen has laid which egg, as well as how many eggs she lays.

However, the effort involved with trap nests is significantly greater since each animal has to be freed from her prison after egg-laying.

In the case of an open nest, just as with a trap nest, bedding litter or, for example, a wooden frame with sisal stretched over it is provided as a base, with a hole in the middle through which the egg can roll away along an inclined board and into a drawer. The litter has the disadvantage that it quickly becomes dirty and therefore has to be changed often. On the other hand, it gives the hen the impression of a natural nest. The more complex roll-off nest gives the hen no opportunity to come into contact with her produce, so that this design prevents egg-eating and also guarantees clean eggs. A further possibility lies in building what is referred to as a "family nest" which, due to its larger space, can be used by several hens at the same time. To avoid broken eggs with this nest type, it is advisable to provide deep litter with sawdust and wood shavings.

Furthermore, the nests should be easily dismantled and should be provided with an angled roof so that the hens cannot settle on them and soil them. A better solution, and one that we ourselves have used, is installing the body of the nests in the equipment room of the henhouse. In this way, we create more space in the henhouse proper and also do not need to walk through the litter to remove the eggs.

Good to know

In order to make it easier for the hen to reach the nest, a landing bar should be provided and the lower quarter of the nest opening should be provided with a board so that litter and eggs cannot fall out.

Good to know

It is important in all cases that the nests are mounted high enough so that hens can also move about easily beneath them and so that pests (e.g. mice) have no chance of building their nests in them.

Feed and Water Containers

For the supply of feed and water to the animals, in principle, the simplest of containers are adequate, for instance, a simple wooden or stone trough for the feed and a preferably earthenware bowl for the water.

More suitable, however – even for small flocks – are the automatic or semi-automatic

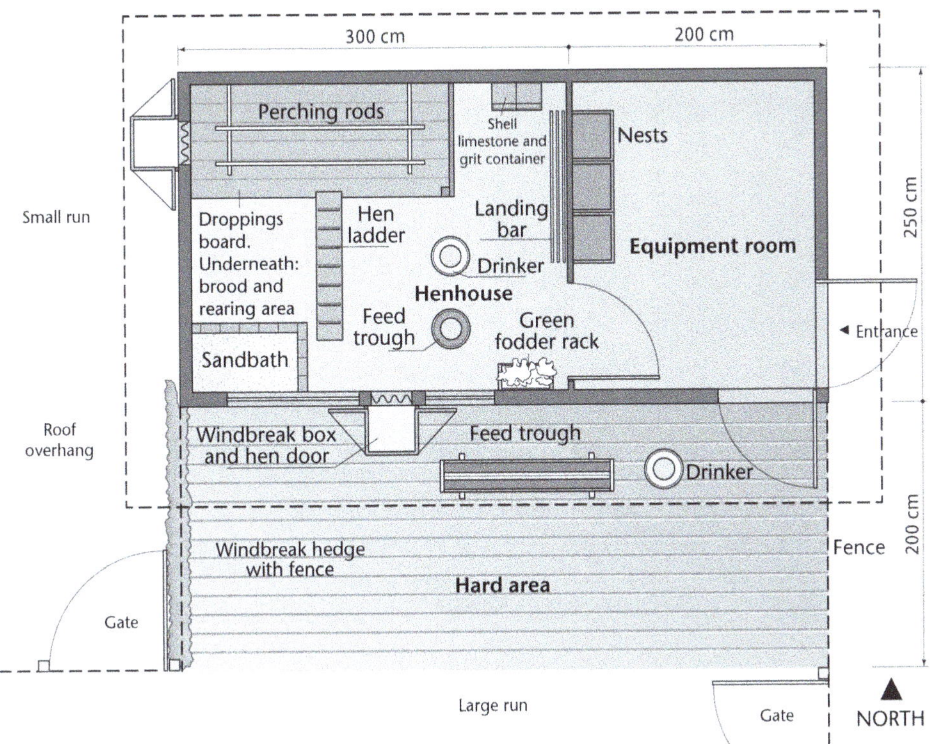

This is how a complete henhouse layout should look.

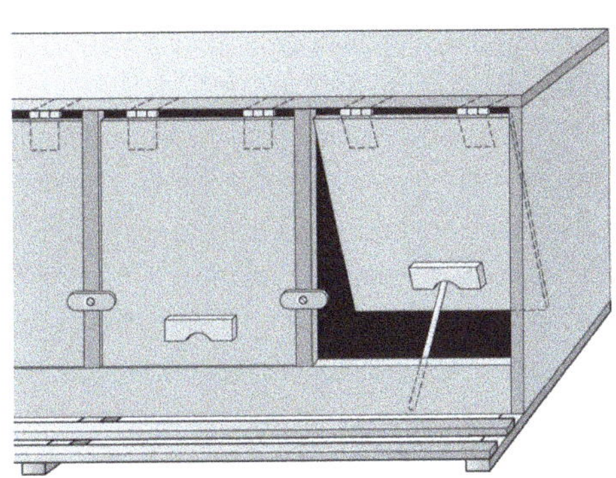

Trap nests allow precise monitoring of the laying performance of the individual hens.

A domestic chicken should be more than satisfied even with such a simple nest.

devices that are commercially available. The greatest problem or annoyance with the aforementioned primitive containers is that in use, the feed or water very quickly becomes dirty, and a considerable amount of the feed is wasted or is consumed by uninvited guests. A container should be provided, at least for the expensive layer meal, and also for water, which saves daily refilling, is easy to clean and allows only little wastage.

These simple bucket drinkers can easily be made at home.

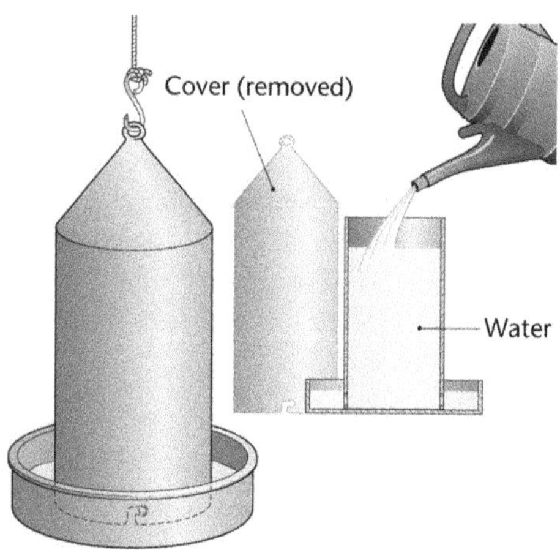

The reservoir drinker is a great labour saver.

Economical and proven devices are the *circular automatic feeder* and the *reservoir-type drinker*. Depending on their capacity, both types of automatic device are sufficient for two to five days, provided they are used in proportion to the number of animals and the size of the henhouse. The automatic feeder should be fastened by a chain to the hut roof and set to a height that is easily reached by the animals or is placed on a firm support construction and preferably filled with feed in the form of meal or pellets. It is adjustable to set the run-out quantity. Care should be taken that the ring-shaped trough channel is only two-thirds filled in each case, so that the birds cannot fling out too much feed with their beaks.

The drink dispenser is similarly fastened to the roof and, like the automatic feeder, is hung suspended over the hut floor at a height so that the hens can comfortably consume the water or feed.

The ring-shaped drinker channel should be rinsed or tipped out daily because dust and litter remnants always settle in the water due to the hens' scratching about.

A great advantage of this supply system deserves to be mentioned. Even the enthusiastic chicken keeper will sometimes want to go away for a weekend or longer. Once such a low-maintenance feeding system has been installed in the henhouse, it will surely always be possible to find someone who can sacrifice a few minutes a day for the chickens. But no fear: this easily automated system will not necessarily cause us to grow apart from our animals.

The daily offering of grains and green fodder maintains close contact with our feathered friends.

For green fodder, a separate container in the form of a hanging basket or a fodder rack fixed to the hut wall are advantageous

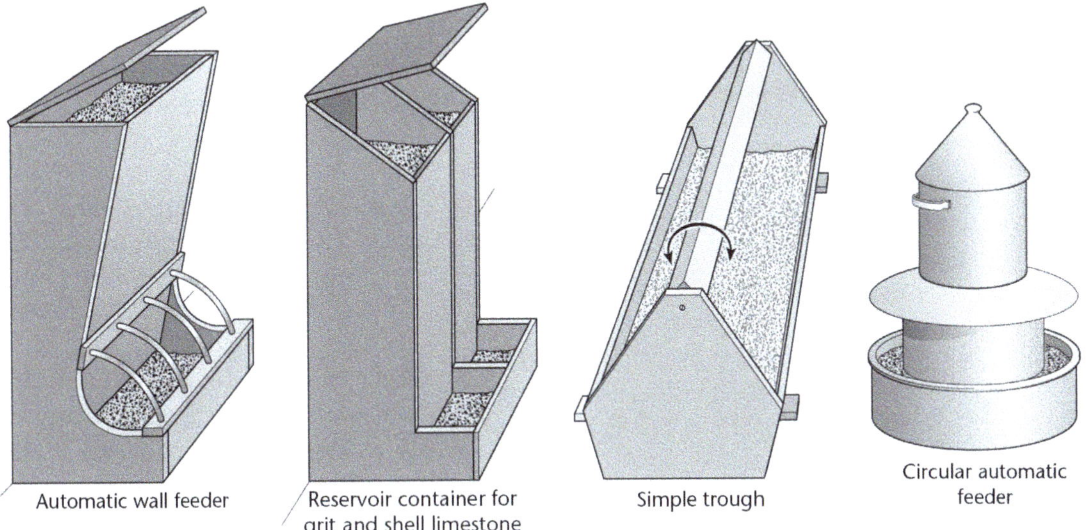

Here is a selection of very suitable feed containers.

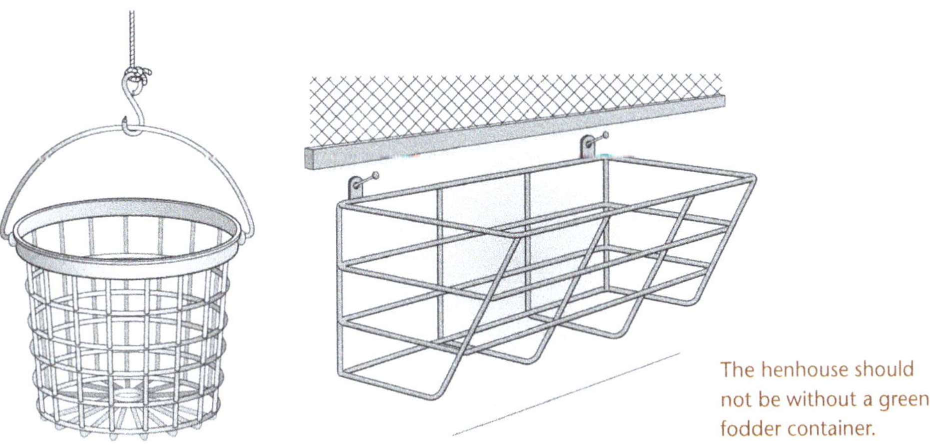

The henhouse should not be without a green fodder container.

because then the green fodder cannot become dirty or be trampled as a consequence of the play instinct. Automatic feed and water dispensers and green fodder containers can be placed over the droppings pit or the scratching area. The scratching area is preferable because feed thrown out of the trough can still be picked up from the floor.

Scratching Area

Watching a hen closely during feeding reveals that its head and feet are constantly in movement. It does not eat everything unselectively, but even with the largely homogeneous laying meal, chooses carefully by size and texture, always testing the structure of the feed. If it moves within the space in front of the droppings pit or under the droppings board in the litter, it constantly wanders about searching and scratching. Because of this behaviour, which is unique and instinctive to the chicken, the littered region is referred to as the scratching area.

The litter material has other important functions, including binding the birds' droppings. Due to the fact that the hen has only one waste exit from its body and no separate urine outlet and that it produces droppings together with urea very frequently (three to six times per hour), this function of the bedding litter is very important to ensure the least possible odour and also that the henhouse stays as dry as possible.

In addition, the soft bedding litter also satisfies a further inbuilt instinctive chicken activity, that of bathing in "sand", which means that they lie in the litter and powder themselves by moving their legs and wings, in order to combat parasites. However, this effect is greatly enhanced if we provide a proper sandbox for the birds, kept separated by wooden boards or stones from the surrounding litter and arranged so that the sun can shine full onto it through the window.

The litter performs the further function of a heat source during the winter – in a sense, a central heating system. If the hut floor is optimally designed – as described in the section on henhouse construction, an active microbial flora develops in the litter, with the result of the conversion of the chicken droppings into compost. This process produces heat, which is largely released upward and warms the henhouse air.

Good to know

Scratching and thereby satisfying a need that is unique to them is something that hens will only do on a soft, yielding surface. For that reason, we should provide them with such a surface.

Bedding Material

Bedding materials that are most suitable under the aforementioned aspects are finely chopped straw, peat moss, coarse sawdust or wood shavings. For cost reasons, a mixture of short chopped straw and shavings are most economic and thoroughly suitable. Sawdust has the disadvantage that it is occasionally eaten by the birds and gives them a feeling of being full without supplying any energy, and therefore naturally has a negative effect on egg and meat production. Peat moss has the drawback that it creates a great deal of dust and is therefore not conducive to good henhouse hygiene.

However, peat moss as a thin layer on the droppings boards or as an occasional addition to the droppings pit has proved useful since it binds the dampness and odour more strongly than anything else.

Absorption capacity per 100 kg of bedding material	
Wood shavings	145 kg
Sawdust	152 kg
Wheat straw	257 kg
Rye straw	265 kg
Oat straw	275 kg
Peat moss	404 kg

The Run

A run, of as large a size as can be managed, that affords the animals plentiful light, air and as much nutrition as possible should adjoin the henhouse. Experts are not entirely agreed about the minimum size for a chicken run. In case of doubt, a small well-cared-for run is, in any event, better than a larger run in a back yard among all kinds of junk and on a side facing away from the sun. As a realistic point of reference, it should be noted that at least 10 m^2 needs to be available per bird. Thus, in

A run in the back yard is not the ideal playground. Apart from the danger of injury from the wide variety of equipment standing around, in many cases the hygiene conditions are poor.

The result of poor pasture care and overgrazing by the hens results after a short time, as here, in a run without any growth at all. But we should not let things get this bad.

an ideal case, the run is large enough so that it can be fenced in at an affordable cost and can be subdivided at least into two parts. This subdivision offers the possibility of moving the animals from a grazed-out and well-scratched pasture to a fresh patch and to repeat this process as often as required.

"Berta", a miniature breed hen strolling in the authors' garden with her offspring over the pasture with its sturdy turf – this is how things are meant to be.

Vegetation

An ideal run does not consist only of a fenced and more or less green area. It is a whole habitat that conforms to the needs of the hens. If one only has a run of restricted size available, particular care must be taken over the vegetation, since otherwise the animals will soon have to search for their pasture on a bare surface, which, in wet weather, quickly turns into a muddy mess.

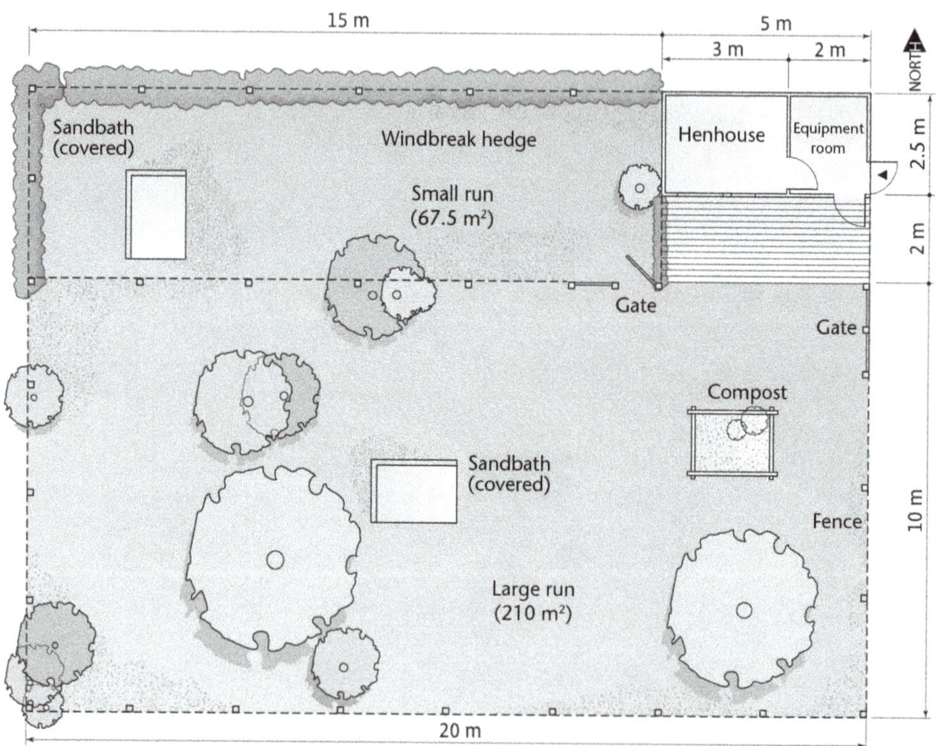

The animals will have a sense of wellbeing in a run of this type.

For sowing, a mixture of perennial (or English) ryegrass, timothy grass, meadow fescue, rough meadow grass, red fescue and crested dog's tail, as well as bird's foot trefoil and white clover has proved to be ideal. This mixture can be made up individually and possibly also enhanced with other grass and clover types, but the bulk of it should consist of ryegrass. This community of different grasses and clover types has the role of protecting the soil and the organisms contained within it and also serves as nourishment for the chickens. The plant cover is subject, particularly on relatively small areas, to very high wear since the chickens constantly work away at it, scratching it with their sharp claws to reach insects, seeds and all kinds of edible things between the foliage and the ground surface.

However, since the chickens do not graze on all the greenery, from time to time small scrub-like islands of withered plants form, which we eventually mow down and distribute over the run, together with any mole hills that have formed.

It is generally worthwhile mulching the whole grass area twice a year and, depending on the soil conditions, to lime it at least once a year. After this, the run should not be grazed for one to two weeks, if possible. Manure is provided by the chickens themselves.

Compost Box

A further valuable feed source and simultaneously a fertiliser for the garden is provided by many types of organic household and garden waste, which are placed in an area enclosed by boards where the hens happily eat all but the inedible parts and rip up the rest. This waste, including fallen fruit, grass mowings, vegetable leaves and other things becomes mixed with the excrement of the animals and is finely torn up by them, providing what is possibly the very best compost, which can be removed now and then and mixed into the normal garden compost. By adding small quantities of lime, a compost with high fertilising power is produced.

A tree protector made of wood or wire is a good idea, so that the tree roots cannot be exposed by the scratching hens.

Shade

If possible, the run should be arranged so that it includes one or more trees or shrubs. Particularly on hot summer days, the animals like to siesta in shade in the middle of the day. A broody with her chicks may well find a great diversity of protein-rich nutrition in the form of insects and worms here. A tree or shrub also enriches the landscape round the house, perhaps providing fruit, and additionally offers the chickens cover from enemies such as birds of prey.

Good to know

If there are no trees and bushes present or if they are still too small to provide shade, the construction of a shade roof to protect the birds on a hot day is a good idea.

Sandbath

The wellbeing of the birds is certainly enhanced by the addition of a covered dust- or sandbath, which should be protected against the wind on two sides. This piece of equipment is very easy to create from four wooden posts with an inclined single-pitched roof and boards enclosing the rear side and the side facing the wind. Under this, a flat hollow is excavated and filled with sand into which, from time to time, some wood ash can be mixed to destroy parasites.

Hard Area

It is worthwhile laying a hard area in front of the henhouse which allows the chickens to get some fresh air and stretch their legs even when the weather is bad. Preferably, this area extends across the whole side of the henhouse facing the run, with a depth from front to back of about 2 or 3 m.

A suitable substrate is coarse gravel or duckboards laid on gravel or a concrete base.

It is important that the substrate allows water to run off rapidly – i.e. is well drained – so that rain or the garden hose can easily wash away the excrement left there by the animals, into the substrate. Particularly on sunny winter days, this part of the run is somewhere the hens particularly love to spend time, provided it is dry and free from snow. If duckboards are used, then care should be taken that they are not dimensioned too large, so that they are easier to clean and, if needed, to repair and replace.

Fencing and Wind Protection

The whole run should be protected by a fence so that enemies of the chickens, where such

On hot days, our animals are glad of a shady spot under trees and bushes. They can cool themselves here and find nutritious morsels to eat.

The arrangement shown here, a solid surface placed in front of the henhouse, is simple and practical. Very worthwhile, also, is the feeding area with a roof over it.

day-time enemies still occur in our suburbs and country areas, cannot get in and take the birds away. Suitable materials for fencing are wire mesh, wooden fencing or solid walls. The most advantageous, though perhaps not the most elegant solution lies in erecting a wire mesh fence with tight-meshed chicken wire in the lower part. It is relatively economical and sufficiently secure, is readily constructed by the DIY enthusiast and does not create shadows, which is an important point when the sun is low during the winter months and particularly for runs with a relatively small ground area. The fence should be between 1.8 and 2 m high since even the highly bred modern-day chicken still has an astonishing ability to fly.

The gate should be dimensioned so that it is possible to pass through comfortably with a wheelbarrow and other equipment. In large runs with, for example, fruit trees, an additional larger gate is worthwhile so that a tractor or a motor vehicle with a trailer for fruit harvesting, fruit tree pruning or for grass mowing can pass through.

If the run is very exposed to the wind, a protective hedge or other windbreak in the form of a solid board fence or a wall should be provided on the side facing the wind. Two or three metres is sufficient to form a relatively wind-free corner in which the animals will shelter during gusty weather.

If one chooses to raise chicks, it should be ensured that the lower part of the fence consists of finely meshed wire up to a height of 50 cm, since otherwise chicks that slip through the coarse-meshed fence and no longer protected by the broody, will quickly fall victim to cats or unattended dogs.

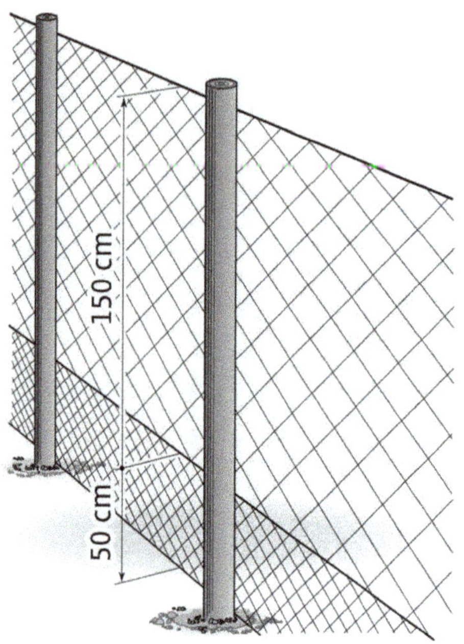

Feed

In the wild, a chicken searches for its own food. It feeds itself primarily on seeds of all kinds, foliage of different types and on a wide variety of animal life, such as insects, worms, beetle larvae and snails among other things. This provides the animal's sustenance needs for its normal life processes. However, we demand more from our livestock. Whereas in its wild state, a hen hatches and brings up one clutch per year, her domestic cousin is required to produce 20 times as many eggs. But since, firstly, we can only make limited feeding space available for the domestic chicken and, secondly, we demand so much from it, we need constantly to offer it high-value nutrition and fresh clean water. On the following pages, we examine what the birds' feed and water requirements look like, how nutritious feed mixtures can be created and what feeding techniques are available and, in particular, useful.

Feed Requirements

Based on the assumption that the animals engage in daily grazing, but not including this intake, for its maintenance energy supply and in order to achieve a satisfactory laying performance, a medium-weight hen needs approximately 120 g of dry feed a day in a mixture that meets the animal's needs as well as possible.

This ration should contain:

- about 60–65% carbohydrates (cereal),
- about 20–25% vegetable protein (soya, pea or rapeseed meal),
- about 5–10% fat (oil cake),
- about 5–10% mineral substances (shell limestone, grit, special preparations),
- about 4–9% milling by-products (bran), and
- up to about 1% vitamins and trace elements (ready-made preparations).

It is, of course, open to every chicken keeper to buy these components individually and to mix them relying on his or her own experience. The percentage amounts shown are not absolute, but should be regarded as a guide. It would quickly become apparent that it is not feasible to react to short-term changes in laying performance with meaningful changes to the mixing ratio of the feeding rations. The external and internal influences bearing upon the complex physiological processes involved in egg formation are too numerous and varied for that.

Feed Preparation

In our experience, it is advisable to use a ready-made quality-assured standard feed mixture in meal form as the basic feed and to enhance this with a proportion of one third cereal or a cereal mixture. Naturally, the supplemented cereal can be omitted and the standard feed mixture used alone. However, this has the disadvantage that the cost is significantly higher since, for the cereal, we can use cheaper waste cereal; that is, broken seeds and cereal mixed with threshing waste.

Another extreme would be to do without any industrial feed mixture and only to feed high-quality cereal, bran, suitable protein sources and specially formulated vitamin preparations. This method is advisable only for those who avoid industrially prepared feed for ideological reasons and have sufficient time to spend on caring for their chickens more intensively.

On this basis, a daily ration might look as follows:

Morning: Soft feed in the form of soaked barley (15 to 20 g), wheat bran (5 to 10 g), soya meal (10 to 15 g), vitamins (1 g).

Afternoon and evening: A total of about 100 g of wheat grains with a modest proportion of maize meal, occasionally skimmed milk curds (equating to a daily ration of 5–7 g). A question that needs considering is that of which feeding methods we should use, both to suit the birds themselves and to achieve success in our chicken keeping.

It is advisable for both purposes, if for example in winter no run use is possible and no greenery is available, to add germinated wheat, barley or oat grains to the feed.

Feeding Technique

Based on our recommendation of feeding two thirds standard feed (laying meal) and one third cereal (wheat, barley, oats, maize), it is most advantageous to offer the laying meal in an automatic feeding device for the birds to take on demand. The commercially available circular automatic feeders, for example, offer the guarantee of fault-free functioning, clean feed and low feed losses.

This observation is helpful, in particular, for the cold and wet times of year when the animals cannot go out into the open every day. At these times, they can become bored in the henhouse and can turn, for example, to bad habits such as feather-pecking or egg-eating. Therefore, the longer the chickens are employed in feeding, the smaller is the risk that they will resort to such vices.

In addition, we also scatter grains for our chickens in the litter and these are quickly picked up and satisfy them so that tiredness and quiet rapidly descend in the flock. With this daily dispensing of grains, we are also able to maintain the necessary close contact with our birds that most chicken keepers want – enticing the chickens by calling to them, while at the same time trying to accustom them to being touched.

If we resort to the intensive feeding method, the effort involved is greater. With this method, pots should be provided in which the soft feed is given to the animals. The vessels must be thoroughly cleaned after use – just like the feeding troughs – since the remains of feed quickly spoil and then lead to digestive disturbances. Vitamin preparations are also available to be bought, like the skimmed milk curds, which can be offered in a separate trough. Since in this case the grain feed mixture represents the main component of the overall feed ration, it should consist of high-quality cereal with a large proportion

Good to know

The meal-like consistency of the feed causes the animals to spend longer with the intake of feed than if the feed is in the form of grains or pellets.

of wheat if the intention is to achieve a correspondingly high output from the birds. Maize should only be given in moderation, otherwise the animals will put on fat.

The provision of grit is advisable, and this can be placed in a combined trough. Grit refers, in this context, to tiny stones, which facilitate digestion and the grinding down of feed in the digestive system. It is especially important to provide grit where the feed has a high proportion of grain and the chickens have little opportunity to go outside. Of course, chickens have no teeth with which to chop up their food. That job has to be done by their gizzard. In order to help it do its work, when they are out in the open, the chickens consume small stones. If, under unfavourable conditions, they fail to find any, we need to make this digestion aid available in the henhouse.

Water Needs

One of the most important but often badly neglected components of feeding is water. The daily water requirement of a hen is about twice as much as its feed intake, so about 250 g per day. The most suitable equipment for small and medium-sized flocks are circular automatic drinkers that are hung at about chicken-breast height in the henhouse.

Even though the water usually lasts for several days in drinkers of this type, we should be in the habit of filling them with fresh water every day or at least swilling out and cleaning the drinking channel daily. For relatively large flocks, nipple-type drinkers are very suitable. These have the advantage that they cannot become dirty, or only slightly, and therefore cause little work. With these drinkers, one nipple should be provided for about every four animals.

CHAPTER 6

Incubation

A dream of many a chicken keeper is to breed chicks from his or her own stock. This is a natural enough wish and also makes economic sense for people with small flocks. And for those who have never had the experience, it is a true delight to watch chicks hatching and to witness the growth and development of chickens from the very start of their life.

Of course, things are very different with chickens than with most other domestic animals. From an everyday, unexciting hen's egg, all of a sudden, new life breaks out into the world. All one needs to accomplish this miracle is a fertilised egg, a place for its incubation, the right temperature, humidity and oxygen supply, also some skill coupled with a little patience, although the last five items in this list can be achieved very easily simply by providing a broody hen.

The Hatching Egg

Essential for success in incubation is faultless conditions both inside and outside the egg. It is also important, of course, that there are sufficient cocks in the flock to ensure reliable fertilisation of eggs. As a rule of thumb, it is usually reckoned for heavy hen breeds, that one cock is needed for every 10 hens, and for light breeds, one cock to 15 hens. It should be ensured that young, vital cocks are included among the flock, since once they pass the age of two years, males are regarded as being "past it" and would be more use in the kitchen, being still tolerably enjoyable to eat at that age.

Further factors that are of decisive significance – equally for both hen and cock with respect to the quality of hatching eggs – are the quality of the feed and water, a well-balanced henhouse climate, optimum lighting conditions and, in addition, the proper collection and storage of carefully selected eggs for incubation.

Some facts and figures on incubation		
Natural incubation		
Duration		21 days
Hatching weight		35–45 g
Chick sex ratio		50:50
Main brood time		April–June
Number per broody hen		13–15
Brood nest size		45 x 45 cm
Artificial incubation		
Temperature	1st–17th day	37.8–38°C
	18th–21st day	37°C
Air humidity	1st–19th day	55–60%
	20th–21st day	80–90%
Turning	1st–17th day	3–4 times daily
Candling		7th and 17th day
Storage duration of hatching eggs		max. 14 days
Storage temperature		12–14°C
Storage air humidity		75%

This broody of ours is very satisfied with this simple wicker basket for incubating her eggs.

Selection and Storage of Eggs for Hatching

For a medium-weight breed, the egg should be about 50 to 60 g in weight, clean, absolutely undamaged and not more than 14 days old.

It is best stored on a wooden or metal storage rack at 12 to 14°C and a relative air humidity of 75% in a well-ventilated but draught-free room.

If one finds oneself in the situation of not having enough "brood-fresh" eggs available for a suitable hatching date, it is possible to freshen up suitable eggs that are older than 14 days by placing them in a pot of warm water at 36 to 38°C. The pot should be insulated so that it maintains the temperature at not more than 38°C as constantly as possible for up to three hours. Within this time, the eggs re-absorb some of the moisture which – due to their relative age – will have evaporated,

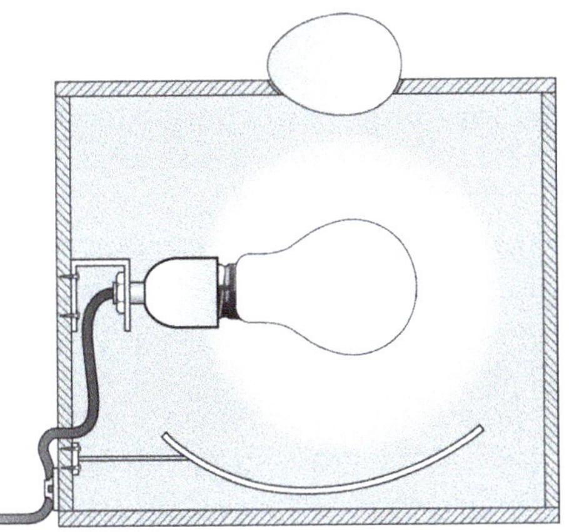

The candling lamp is an indispensable aid for checking hatching eggs, as well as eggs intended for consumption.

and they become hatchable again. But this method can be no more than an aid and will normally succeed only when combined with natural incubation. Freshened-up eggs will fail under artificial incubation in most cases.

Before being placed in incubation, the eggs should be checked with a candling lamp for fine hairline cracks, blood spots and, most particularly, the correct position of the air bubble (at the rounded end). All three factors are essential for incubation success and are not externally visible.

Structure of the Fertilised Egg

Before we consider the incubation process itself, it is interesting to remind ourselves just what an amazing thing it is that we are dealing with. Embedded in the egg white, or albumen, is the yolk, which consists of three different layers surrounded by the yolk membrane. Inside the yolk itself, a structure called the latebra carries the germinal disk with its germinal vesicle and is structured so that it points upward whatever the position of the egg.

The egg white surrounding the yolk with its blastoderm also consists of a number of layers, of which the first is a thick albumen layer which forms the chalazae that are anchored at the two poles of the egg and hold the yolk in position, floating in its protective layer, but still able to rotate around the egg's longitudinal axis. Following the first thick albumen layer are a thin and then a thick layer and then a further thin layer, which finally is enclosed by the inner shell membrane.

The outer covering of the egg consists of the outer shell membrane, the shell itself as we normally see it, and then the outer skin or cuticle, which gives the egg its shiny appearance.

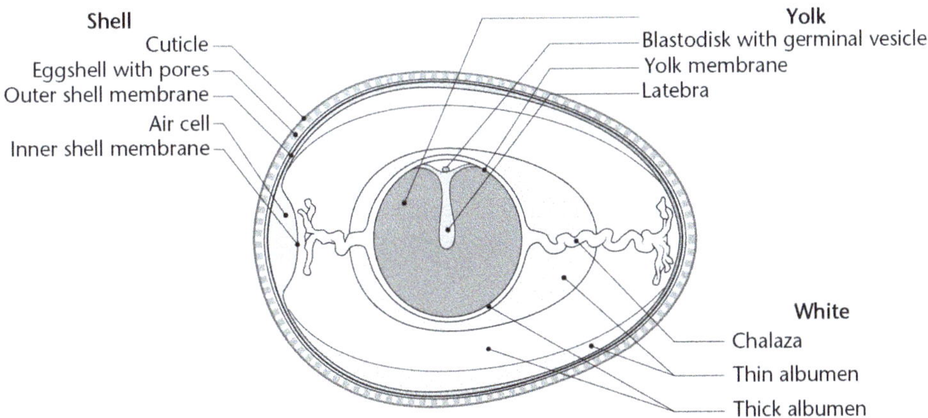

The egg in all its detail.

The structure of the shell consists of a net-like interwoven organic substance with a filling of an inorganic mass, the latter being mainly calcium carbonate. Although the shell hermetically seals the egg against foreign bodies with its internal membrane and outer cuticle, thanks to around 10,000 tiny pores, it ensures the indispensable gas exchange between the developing chick and the outside world. Despite these, it has astonishing strength and also meets the need for the tender chick to be able to break out of the shell at the correct place when the time is right.

Incubation times compared	
Pigeon	18 days
Chicken	21 days
Turkey	27 days
Duck	27 days
Goose	30 days

Incubation Methods

A distinction needs to be drawn here between *natural* and *artificial* incubation, both of which are possible and suitable for small flocks. However, natural incubation requires more instinct and patience from the keeper, since success depends to a large degree on the broody hen and on the factors in the surroundings which affect her. If the chicken keeper places more value on reliability and therefore on economic viability, then the artificial method is certainly to be recommended. The word "artificial" is not used here in a derogatory sense, but only to indicate that the incubating environment created by a broody is mimicked by a man-made device.

Incidentally, artificial incubation techniques have been used for centuries in some civilisations, as revealed by evidence from ancient China and Egypt. In Europe also, artificial incubation of chicken eggs has been practised for around 200 years. In the older techniques, heat sources have been provided by burning wood, hot water systems and even horse manure. Unsurprisingly, these methods were highly unreliable, but welcome nevertheless as alternatives to natural incubation and, over time, they evolved into the forerunners of modern commercial chicken production.

With the coming of electric power, humanity gained the possibility to develop a less labour-intensive and more easily controlled system which guaranteed a high level of hatching success. It was only then that the basis was created for the mass breeding and rearing of chickens.

Natural Incubation

Natural incubation by a broody hen is unequalled in the excitement it offers for the enthusiastic keeper of a small flock. And who would not be enraptured at the sight of a hen with her tiny feather-ball chicks, a wonder that very few people have the chance to experience nowadays?

The Broody Hen

Right at the start, the question arises as to how we might find a broody hen or how we can tell if our own hens will brood. In many miniature breeds and old country varieties, we are unlikely to have difficulties in this direction. When the days become longer and warmer, if there is fresh green fodder and other delicacies available in abundance, we might well notice that one of the hens suddenly starts behaving oddly and that her appearance starts to change.

She emits clucking sounds, separates herself and roams around searchingly, develops a shrunken, whitish-coloured comb and begins laying. She also avoids the cock, and if we approach her nest, remains firmly in place on it. If we take her out of her nest despite her threats or even fierce defensive actions and discover on her belly the familiar brood patches – reddened areas on dry, almost featherless skin – our suspicion is confirmed. If we pay no attention, one day she might be temporarily absent and brood in a quiet corner or settle on the eggs of another hen.

The reason for this altered behaviour is a change in her hormone balance. The physiological circumstances which determine brooding behaviour are still not known, however. In the past, it was thought that when becoming broody, an increase in a hen's body temperature took place in the region of the breast and belly, which was supposedly uncomfortable for the hen, and that she would try to counteract this heat through contact with as many cool eggs as possible.

When, in the course of time, the temperature of the clutch reached her body temperature, she would turn the eggs with her beak in order to move the cooler undersides to the top. And that was how the mystery of why a broody hen will sit on a clutch of eggs for so long was explained.

This reasoning might have seemed perfectly obvious to a lay person and it does also fit with actual observations, but nowadays we know that the body temperature of a broody hen does not rise either overall or in particular body areas. As is often the case, we are now richer in knowledge, but at the same time, we are poorer for the lack of an appealing explanation.

Good to know

In modern factory farming breeds, the brooding drive has been largely bred out so as not to interrupt egg production; after all, sitting hens do not lay eggs. For that reason, if we keep these breeds, we might well find ourselves waiting in vain for a hen to turn broody.

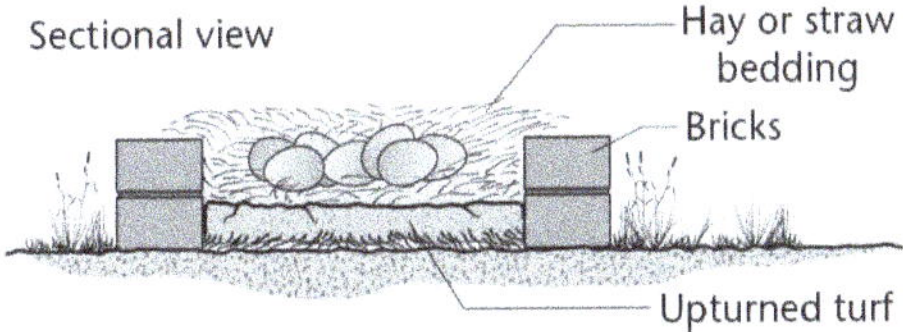

It is easy to create a brood nest, using simple means, on the henhouse floor.

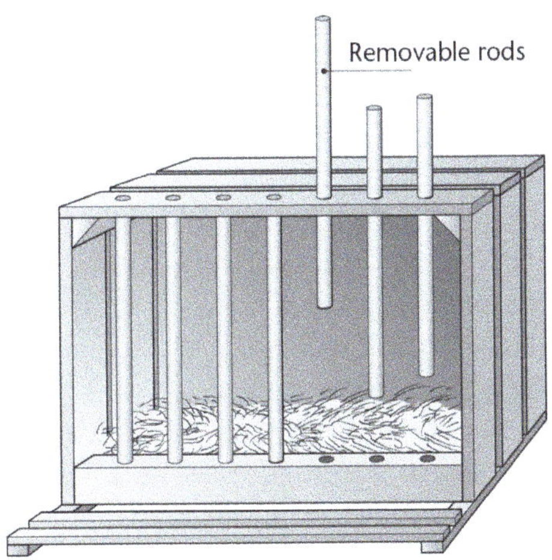

This simple portable brood box can also be used for breaking hens of unwanted broodiness.

Good to know

If we have a medium-weight hen, we can put 12 to 15 carefully selected and stored hatching eggs under her. We can confidently leave everything else to the hen for the next 21 days.

The Nest

It is best to provide the broody with a nest in a dimly lit, draught-free corner of the hen-house, somewhat screened off from the rest of the flock. This can be done by constructing a special nest box which can then be equipped with a softly padded lining. A simpler possibility consists in preparing a nest on the floor. For this purpose, a suitably sized slab of turf can be cut out and laid, with the roots uppermost, on the floor of the hut and this can then be covered over with hay or straw as the nesting material. To prevent the nest being displaced or trampled out, the turf can be surrounded on all sides with bricks.

The Incubation Process

In order to accustom the hen to the nest, it is best, at first, to place some porcelain eggs into it. At the same time, we should try to gather suitable hatching eggs. We can take our time in our search because it will take a few days until the broody-to-be has accepted the nest and is firmly in place brooding.

It is also advisable to treat the hen against parasites before she starts sitting, particularly against the very common chicken mites. If these pests are unable to trouble the hen, she will sit much more quietly on her eggs. And the freshly hatched chicks will also benefit if they are not pestered by parasites right from their first day of life.

If, in the meantime, the hen leaves the nest – usually at the same time each day – we should not be too concerned. On the one hand, it does the eggs good to cool down a little for a while and to be exposed to some fresh air, while on the other, the hen naturally also needs to consume food and fresh water during her demanding brooding period. It would be best to provide a special ration of feed close to the nest for her to help herself as she wants. However, soft food and green fodder should be avoided due to the risk that it could rapidly become foul, and also to reduce the chance of any digestive trouble and diarrhoea. Access to a dustbath should also be provided.

Artificial Incubation

For artificial incubation we need either an incubating device or access to a commercial hatchery located somewhere nearby, to whom we can entrust our eggs.

The Incubating Device

Nowadays there are handy incubating devices even for the small chicken keeper, and at reasonable prices. Some skilled individuals even build their own equipment for this purpose. In a simple but adequate configuration, it consists of a housing with a viewing window, has a metal grid as a floor and is heated by an electric heater that is held at the relevant temperature by a regulator, and also has a temperature and air humidity measuring device, as well as a device that ensures an adequate supply of oxygen.

Depending on the size and construction, the capacity of these small incubators differs. As a general rule, it is advisable, however, not to use up all of the existing capacity of the device. Once we have got the desired number of hatching eggs together, they are placed in the device at the same time and the start date is noted.

Candling

During the incubation period, which lasts a total of 21 days, the eggs are examined for the first time on the seventh day of incubation by shining a light through them; in the specialist jargon, they are "candled".

To do this, we make use of the candling lamp, which we have already used once before to inspect the eggs selected for hatching. Eggs that show no change compared with the day of their placement in the incubator and look clear when candled, are removed since they have either not been fertilised or have died. Fertilised eggs can be recognised at this stage by "spider veins", indicating the formation of the blood vessels. Dead eggs are not easily identified by a lay person at this stage. It is therefore advisable to remove them at a later time. The second candling procedure is best carried out on the 17th day, at which point the temperature should be reduced somewhat and the relative air humidity should be increased.

Helpful tip

A suitable incubator offers the chance to observe incubation and hatching more closely than would be possible with natural incubation. Particularly for children, this is an especially exciting and interesting experience.

The selected hatching eggs are carefully placed into the tray.

The progress of the incubation can be easily observed with this incubator.

Turning

In large commercial automatic incubators, in addition to the regulation of temperature, air humidity and oxygen supply, turning is also carried out automatically. In small incubators, this operation needs to be done by hand.

Three to four times per day is sufficient. In any event, it is essential to treat the tender developing life with care and turn the eggs cautiously by ¼ to ⅓ of a turn about their longitudinal axis so that the embryo does not adhere to the shell.

It is also advisable to mark the eggs accordingly so that we can always be sure of the starting position. The critical phases lie between the third and fifth day when the respiratory system develops and the metabolism undergoes a change to enable the complex processing of protein and fat, and then again between the 18th and 20th days when the transition to breathing with the lungs takes place. From this time point on, the chick produces significantly more internal heat. In order to protect it against overheating and the consequences arising therefrom, the temperature must be reduced by about 1°C.

The Development of the Embryo

Cell division begins immediately after fertilisation while the egg is still in the hen's oviduct and thus long before the start of incubation; it continues until the stage of about 250 cells and then the division process is interrupted, and will only be restarted once the incubation conditions have been established.

If we were to examine the incubated eggs daily, we could closely follow the development of the embryo. Starting from a single cell layer, multilayered structures form, which soon develop into differentiated cell formations and result in the germ layers, specifically:

- the ectoderm or outer germ layer, from which the outer parts of the chicken, like the skin, feathers, feet, the back and the nervous system are formed,
- the mesoderm or middle germ layer, from which the reproductive organs, the muscles and the bones originate, as well as the blood, and
- the endoderm or inner germ layer, which is responsible for the formation of the respiratory organs, the glands and the digestive organs.

All of this takes place before our very eyes in a constricted space enclosed by a brittle capsule in a total of 21 days.

On the 16th and 17th days, the embryo breaks through the closed circulation in the egg. Then the beak pushes into the air cell and the chick starts to breathe with its own lungs. This ends an important phase of development.

Hatching

In the normal position, the little creature is oriented with its upper body toward the air cell and with its feet lying close against the body, also pointing toward the air cell, whilst the head is initially tucked under the right wing. When the hatching process starts, the chick pulls its head out from under the wing and begins using a horny projection, the so-called egg tooth, to break open the shell from inside. At the same time, it heaves its powerful legs in the opposite direction, down and forward, and thus provides the necessary reinforcement for the work with its head. From time to time it pauses to rest from the hard work.

Writing about this in 1876, the German ornithologist Dr Eduard Baldamus observed:

> *For the purpose of facilitating the arduous task of freeing itself, the chick has a sharp and somewhat hard outgrowth on the tip of its upper beak by which means, during intervals of varying length – sometimes 10 minutes or longer – it executes two, three or more powerful blows against the shell. Although the latter is made friable, its fracturing demands a great deal of strength, which appears remarkable when one considers the inconvenient position of the chick, making withdrawal to any distance for the blows impossible.*

Once the chick has finally won the fight and struggled free, at first it seems to be still gummed up and quite out of proportion.

Hatching from the confined egg is exhausting work for the tiny creature. The little chick fights its way with unimaginable energy out of its dark prison and into the light.

But in a short time, it will be standing stably on its powerful little legs and, as soon as it has dried out, it will take on the appearance of a little ball of feathers, to the delight of everyone, and especially of children. And soon, under the guidance of its mother, it will set off on a voyage of discovery.

Good to know

We can, however, influence the environment for the sitting hens by ensuring it is as quiet as possible for them, the henhouse conditions remain as constant as possible and that they have sufficient good-quality feed and fresh water.

Problems with Incubation and Hatching

Things do not always go so smoothly. Some chicks do not manage the last step into the open and remain stuck in the egg, are misshapen or have such a weak constitution that they do not survive the first few minutes and hours.

There are many possible causes for this which we can influence or even completely eliminate, particularly in the case of artificial incubation.

Individual symptoms and the related causes in this list also apply in the natural incubation process. Particularly worth mentioning in this regard are the factors which have their origins in the time before incubation; these are faulty feeding and housing of the parent animals, improper storage of the hatching eggs, and similar causes. It is also important not to place too many eggs under the broody hen; in this regard, less is more. But as far as the incubation factors themselves are concerned, there is little we can do, and the behaviour of the hen itself is decisive for hatching success.

Unwanted Broodiness

As difficult as it often is nowadays to find suitable animals for natural incubation, the opposite problem can also arise that, if the weather is good, we find ourselves with too many broody females in the henhouse. This means fewer eggs and a great deal of unrest in the flock. Again, humans have often

Causes of poor hatching results	
Symptoms	Possible reasons
Eggs with no signs of development	Too few or even too many cocks; hens that are too old; flock in poor condition; hatching eggs too old; hatching eggs stored at too low a temperature (applies for natural incubation also)
Dead eggs or embryos after first candling	Specified incubation temperature not maintained (too high or too low), insufficient oxygen supply; temperature variations too great (particularly cooling), errors during turning (too little, too often)
Pecked eggs with chicks dead in the shell	Insufficient moisture, temperature too low or brief but severe temperature rise
Moist and gummed-up chick sticking to shell	Dried out due to too little moisture during hatching
Chick wet and gummed up with contents	Temperature too low and air humidity level too high during incubation
Malformed chicks	Largely genetically caused, but also errors during turning or wrong incubation temperature
Dead chicks with bad smell	Navel infection (due to poor hygiene in the incubator)
Undersized chicks	Eggs too small (applies also to natural incubation), incubation temperature too high and air humidity too low
Large, weak chicks	Incubation temperature too low, air humidity too high and lack of ventilation
Premature hatching	Incubation temperature too high
Late hatching	Incubation temperature too low

developed methods – some of which can only be classed as cruelty to animals – both for inducing hens to become broody and for dissuading them from broodiness. Particularly the means for preventing broodiness have found expression in some unusual ways. For example, the animals were – and are – held for seconds at a time head down in ice-cold water, stuffed into sacks, placed for days in cold cellars or put into hot ovens.

A more humane technique from a previous age was described in a passage written in 1892 in a German language book on poultry by K. Römer; this reads as follows:

The most suitable method is to place the broody hens into a cage or a similarly constructed basket which is placed in the open air in the chicken run. As a result of the captivity, as well as the constant presence near the cage of the other poultry, particularly the cocks, the broody hen is kept in a constant state of agitation, no longer sits and within a few days has forgotten about her broodiness. Immersing broody hens in cold water is pointless and easily leads to colds; it is preferable to place a deep bowl with water in the hen's brooding place and to cover it over with a thin layer of straw, so that when the hen sits down, she enters the water, thus inducing her to leave the nest.

The only one of the tips and tricks mentioned that seems to us to be worth mentioning and which promises success is the withdrawal cage. However, the wire cage described in relation to rearing chicks would be just as useful.

Breeding Planning

Before turning to the topic of rearing chicks, it seems appropriate here to consider which animals to bring on, and in what form; that is, selecting the hens that we wish to include in the flock as efficient layers and deciding which ones should be separated out immediately for the table. It would be sensible to make this selection and corresponding pre-selection by the time that the decision is made concerning which eggs from which hens should be selected for hatching.

Observing this simple breeding rule should probably be sufficient for the small chicken keeper who does not have any great plans regarding the output of his or her flock. However, for the individual who does have the ambition to improve his output in a specific way and who combines the joy of chicken keeping with raising particular pure breeds, the results obtained with this simple method will be too vague and, more significantly, will take too long. After all, this individual breeds not only for yield (eggs, meat), but most particularly for a defined appearance precisely laid down for a specific breed. This itself covers a wealth of features, such as body shape, plumage characteristics, coloration and more.

All these external features and the yield features (laying output, egg size, fertility, vitality, feed conversion ratio and others) are genetic; that is, they are pre-determined by the hereditary factors and are combined anew for the offspring from the characteristics of the parent animals at fertilisation. How this combination of hereditary factors takes place, whether some features cancel each other out or, conversely, become reinforced takes place according to the precisely defined set of genetic principles discovered by Gregor Mendel.

Mendel famously crossed red flowering peas with white flowering peas and obtained pink flowering offspring. When he crossed these with each other, he found that a quarter of the offspring had pink flowers. Finally, when he crossed red or white flowering plants with pink flowering individuals, half of the offspring were either red or white flowering and the other half pink. However, crossing red flowered plants with one another always produced red flowering plants in the absence of any genetic segregation; the same applied for the white flowering plants. The fundamentals of inheritance outlined here relate, in the experiments carried out by Gregor Mendel only to a noticeable characteristic, specifically the colour of the flower petals. In the breeding of chickens, there is – as mentioned before –

Good to know

The basic rules for productive further development of the flock lies in always only using the best laying hens for increasing the quantity of livestock and not cutting corners when it comes to keeping a good breeding cock.

Good to know

In the event that inbreeding occurs continuously, there is a danger that it could have a negative influence on other features like vitality and fertility. At this point, it is advisable to stop and introduce some fresh blood from outside the flock; after all, what use is the most beautiful hen if it cannot reproduce?

a whole series of features to observe and to influence.

To cover the methods in detail would fill another whole book, but the following principles should be noted: the breeder makes use of the genetic inheritance principle that some features are clearly passed on *homozygously* from one generation to the next. The homozygosity of a feature can be assumed with a high degree of probability if the coloration within the breeding flock or the breeding stock is inherited without any variations. As breeders, we can achieve this by always pairing animals that have the same feature.

In contrast to human society, inbreeding – the pairing of close relatives – is not a moral impediment, but rather a tried and tested, highly successful breeding method for reaching a goal rapidly and reliably. In fact, incestuous breeding in every combination is not only allowed but is highly successful if it is wished to breed one or more features consistently into a flock. So, to summarise these points: homozygosity, inbreeding and strict selection are the guarantors of breeding success.

Even the non-breeding chicken owner would do well to take note of a few points on the building up and enhancement of his or her flock. When buying one's first animals, it is advisable to go in the company of an expert so as to obtain the healthiest and fastest-growing animals possible, which are also free from physical defects. When breeding using natural methods, one must ultimately take care that, as far as possible, early hatchings can take place, since they usually result in birds that have greater vigour and also begin to lay in the same year. In addition, if we watch our animals a little, we soon discover which ones among them are the best performers. If possible, we should only use the eggs from these individuals for breeding on in the spring and then follow the same procedure again with the offspring, whether hen or cock. Particular attention should always be paid to the selection of the cock, since he passes on his good properties more rapidly than a hen, since with his sperm, he is able to pass on desirable features multiple times unaltered.

In this way, we are able to influence and shape our flock individually through breeding, according to our personal taste and ambition. However, we can also let things take their own course and simply enjoy whatever it is that random luck turns up for us. They will be chickens in any case.

CHAPTER 7

Rearing Chicks

For "artificial" or "natural" rearing, essentially the same applies as set out above for "artificial" or "natural" incubation. Both possibilities can be practised alongside or in combination with one another. For example, artificially incubated chicks can be placed with a hen who already has chicks of her own and it is equally possible for naturally incubated chicks to be raised artificially if, for example, the mother hen has been taken by a fox. The point is that, in each individual case and depending on the actual circumstances and the wishes of the chicken keeper, it can be decided which system is the most advantageous.

Although it is the case with human children that if they lack love and care from other humans, they become underdeveloped both physically and emotionally, and may even die, these fearful consequences are not found in chicks that lack parents. Since the young chicks – as nidifugous birds – once hatched, are already highly developed and seek feed and water themselves, are able to move rapidly and are otherwise much more independent than a newborn human child, they clearly do not need such intensive physical and emotional care. Artificially reared chicks grow just as rapidly as their naturally raised peers and also show the same patterns of behaviour. With proper care, the mortality rates are lower even than is the case with natural rearing.

But anyone who keeps chickens for reasons other than the pursuit of a serious commercial interest should try not to miss the opportunity of watching the natural rearing of little chicks.

Natural Rearing

During natural rearing, when the first chicks have hatched, the nest should be cleared of egg shells and any dead chicks and, if there is expected to be a long delay until the last chicks have hatched, the early hatchers should be put in a warm place so that the broody hen can incubate the latecomers to hatching. A warm place is regarded as having a temperature of 32°C and a relative air

Some facts and figures on rearing chicks	
Space required	10–12 chicks per m^2
Room temperature	18–20°C
Temperature under the heat source	30–32°C (1st week of life)
Air humidity	60–70%
Loss rate	15–20%

humidity of 60–70%. This is roughly the temperature and the climate that the hen must create under her plumage to give her brood a good chance of survival.

When all the chicks have all hatched, the nest should be cleared once more and the previously removed chicks put back with the hen. If there have been many losses during incubation, we can put a number of stranger chicks under her. The best time to do this is at dusk when the hen and her own chicks have already gone to rest for the night.

Henhouse Conditions

The hut should be light, free from vermin, dry and draught free and should be at room temperature of about 18–20°C.

The best litter materials for bedding are sawdust or short chopped straw. Sawdust should not be too fine because otherwise it might well be eaten as feed by the chicks, giving them a sensation of being full but without providing them with essential energy.

For the first few days, the hen and the chicks should be kept separate from the older birds in the henhouse, firstly to prevent parasite infestation and secondly because the chicks could be attacked by the other hens or the cock. A good broody will try to protect them against this, but it is certainly no error to wait until the chicks are nimble enough to flee to the broody themselves and the hen and chicks can fit without difficulty into the main flock.

Running Free

The broody will happily wander about in the open with her chicks and show them the things that are both nutritious and tasty among the rich offerings of the natural world: seeds, insects and worms. Everything is initially strange, frightening and attractive all at once to the small chicks.

Good to know

If there is a long delay in hatching or if the hen is impatient, the ready-hatched chicks should be temporarily taken to a warm, dry place to be brought back later, so as not to distract the hen too much from her incubating job.

The little chicks readily explore their new unfamiliar world.

Good to know

Before the end of their first week of life, the hen and the chicks should not be allowed into the open, nor during wet or windy weather.

It is advisable to dedicate a separate area for the hen and her young ones away from the other chickens during the chicks' early days so that she can give her undivided attention to her brood. Care should be taken that the area is not contaminated with droppings from the older birds or other animals, due to the risk of infections.

If the run is of almost unlimited size, this measure is not necessary, provided the hen does a good job of mothering, since the rest of the birds will be widely distributed and busy with other things. As far as hygiene is concerned, the problem practically solves itself through the much lower occupation density. As a rule, it can be said that the outside temperature should be approximately the same as the henhouse temperature (18°C) or higher. The little chicks should also not be let out into the open before the sun has cleared the grass of dew. For us, it is always an uplifting and moving experience to watch the chicks making their first journeys of discovery.

If it can be expected that the chicks would easily fall prey to predatory animals or birds of prey, then it is advisable to construct or obtain a run covered over with wire and with weatherproof housing into which the birds can flee if the weather suddenly changes for the worse.

The wire-enclosed part should have a ground area of about 2 x 2 m and should be about 50 cm high. Like the housing, it should be made of light materials and provided with carrying handles so that it can easily be moved. It also makes sense, in areas that are severely exposed to the wind, to stretch some plastic sheeting over one side of the wire cage.

Feed and Water

We feed our own small chicks with ready-made special compressed feed or chick meal and finely chopped greenery, such as stinging nettles, dandelion and vegetables; also lumps of turf, earth free from chemicals, finely chopped boiled eggs, some fine shell limestone and, as special delicacies, ant pupae of the black garden ant, which is a pest in the garden. In addition, we provide fresh water in bell drinkers or in a bowl, in the middle of which we place a big stone, so that the chicks cannot drown if they fall in. We provide feed during the first days of life, five or six times a day, on feeding boards or in flat containers which the chicks can reach without great effort. It is particularly important that at each meal time, we only give as much feed as the chicks can consume in a few minutes, since mouldy feed remains easily lead to disease and fatalities.

Average daily feed requirements	
1st–4th week	10–30 g
4th–8th week	30–55 g
8th–12th week	55–75 g
12th–16th week	75–90 g
16th–20th week	90–100 g
From 20th week	100–120 g

For those who prefer not to use ready-made feed for the chicks, a mixture of finely crushed wheat and fine oat flakes with the delicious added ingredients already mentioned above is recommended. In addition, from the second week on, low-fat curds can be given; but care should be taken that the feeding containers are scrupulously clean, so that the curds do not turn sour. Otherwise, diarrhoea and associated deaths can be expected. The hen should be fed separately as far as possible so that she does not take all the chick feed or soil it.

Average weekly feed consumption from chick to laying hen	
1st week of life – weekly multi-use	45 g 45 g
10th week of life – weekly multi-use	450 g 45 g
22nd week of life	800 g

The Role of the Mother

Finally, a word or two more about the mother hen. As with most animals and also humans, the maternal instinct makes a hen unimaginably fierce. Whereas some birds – like, for example, the lapwing – use a variety of deceptive manoeuvres in order to deflect an

Not all the chicks have hatched yet. The hen waits patiently for the latecomers. Finally, they are all there. Trusting and inquisitive, all gather round the hen and dare to make their first forays into the unknown. Now the hen needs to be extremely attentive to keep the exuberant young ones in check and to spot dangers at an early stage.

approaching enemy from their young, the hen relies on attack and spirited defence. Although hens often quarrel between themselves and attack each other, they normally never attack a larger animal or even a human.

Not so the mother hen. If we come even slightly too close to the chicks, she immediately calls them to her and adopts a defensive posture. If we are bolder still, her neck feathers ruffle up and her posture, as she gets ready to jump up, seems to say, "Watch out, an attack is imminent."

Even if there is no danger, the mother's behaviour differs greatly from that of her egg-laying sisters. Tenderly and with a familiar "cluck, cluck, cluck. . ." she leads her little brood through the run. She takes care that the chicks do not walk into wet grass or through puddles, and on hot days, leads them to a shaded place and draws the attention of hungry beaks to ever more tasty morsels. She also makes sure that the little ones are not over-exerted by the long footmarch and corrals them all safely together in the henhouse at the right time in the afternoon.

Fostering

If a misfortune should occur and a mother hen be lost, we need to see if we can find a foster mother who will take on the care of the orphaned chicks. This might be another hen who already has chicks, or even a turkey with chicks. Turkey mothers are generally better at raising chicks and more self-sacrificing than hens. For that reason, in earlier times, turkeys were often kept for the purpose of rearing chicks, particularly since it was possible to put significantly more eggs under them. Turkeys also had the advantage that they could be persuaded to sit on eggs within just a few days, using a special brood box. In this way, it was possible to remain, to a certain extent, independent of the main brood time of the hens. A particularly special creature in this regard is the broody male turkey. He is an extraordinarily careful father and fierce defender of his chicks, such that even large dogs and humans have to be wary of him. This naturally carries certain risks, for example, when the concerned chicken keeper wishes to check the young for diseases.

Artificial Rearing

The essential feature of artificial rearing lies in the fact that the broody hen is replaced with another source of heat. Heat, after all, is the decisive environmental factor for the survival of the chicks, particularly in the first weeks of life when their plumage is not yet formed. Wet, cold weather and draughts lead inevitably to colds and usually to death.

Heat Sources

The best management and the greatest rearing success are achieved nowadays with the electrically powered devices that are now universally used, such as infrared lamps and radiant heaters. The size of the brood has to be matched to the heat output as specified by the manufacturer. It is important that, using a suitable device for an area that includes the whole brood in a loose group, a constant temperature of 32°C is provided, at an air humidity of 60–70% for the first week of life. The ambient temperature in the henhouse should be maintained at 18–20°C. In each further week of the chicks' lives, the temperature of the heating device can then be reduced by 2°C until the henhouse temperature or a corresponding outside temperature has been reached. The maintenance of the values given is no problem given the present state of the art.

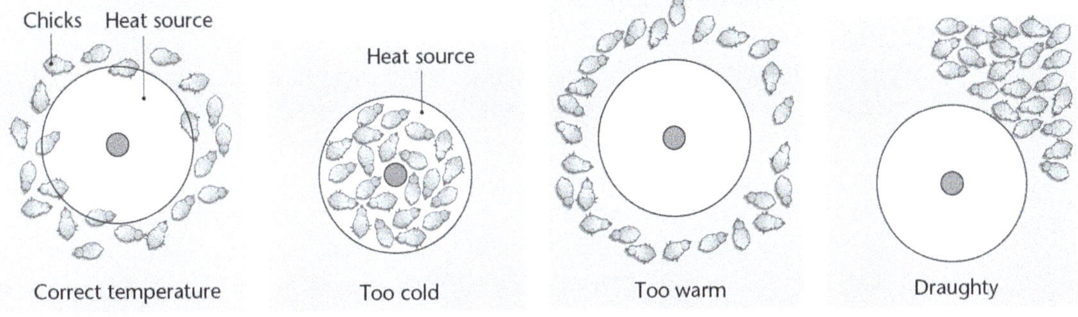

From the behaviour of the chicks under the heat source, we can tell whether the device is correctly adjusted or not.

We can tell whether the temperature as given by the manufacturer is really being maintained by observing the chicks. If the temperature is too high, they will avoid remaining under the brooder lamp and stay at the periphery of the range of effect of the heat source; see the illustrations above.

If they receive too little heat, they huddle together in a clump under the heat source. If the temperature is comfortable for them, they remain in a loose group under the heater and here and there make little forays into the edge regions of their area. In earlier times, artificial rearing used to be more difficult when attempts were made to generate the rearing temperature by more primitive means.

Starting with a packed bedding of horse manure or heating systems that were powered by hot water bottles, progressing to coke stoves, briquette heating or other stoves powered by solid fuels and fitted with a brooding hood, through to gas-fired radiators, everything was tried and practised to make it possible for a larger number of chicks to be warmed at once than a broody hen could ever manage.

The right temperature
Under the heat source, the temperature about 4 cm over the floor must be:
32°C during the 1st week
30°C during the 2nd week
28°C during the 3rd week
25°C during the 4th week
22°C during the subsequent weeks

The Brooding Area

Chick brooding rings have proved to be ideal for segregating the brooding area. In the simplest case, these consist of cardboard strips which are bent into a ring and divide off some

When the mother hen entices her chicks to come out into the open, her brood follows hard on her heels. But the golden rule for them is not to lose contact.

of the henhouse space for their first weeks of life. This is necessary so that the chicks do not go astray and also in order to concentrate the radiant heat emitted by the artificial heat source as much as possible and therefore to use it more effectively. As a rule of thumb, it is usual to allow 1 m^2 for about 15 chicks.

An important factor in the segregation of the space is that it should have no corners, because when they sense danger, chicks easily become panicked and flee in large numbers into a corner. In the process, some of them can be smothered to death.

Litter

The best litter materials for bedding are short chopped straw, coarse sawdust, wood shavings or a combination of these. What is important is that the litter is in perfect condition; that is, dry and clean. It should be regularly refreshed around water dispensers and feed boards or troughs to prevent the spread of disease-causing microbes and the consequences of rotting feed remains.

Feed and Water

The same principles apply generally here as for natural rearing, although it should be emphasised that it is advisable always to provide only small quantities in short intervals so that the feed remains as fresh and clean as possible, because the success of our efforts can all too easily be ruined by dirty feed and water.

The Run

Whenever possible, we should let even motherless chicks go outside when the weather is fine – although not before the 8th day after hatching. Light and fresh air undoubtedly promote both the chicks' growth and hardiness.

The best thing is to allow the chicks direct access to the run from the henhouse, where possible separated from the adult birds so that the young will not be maltreated by them and so that they will not come into contact with the droppings of the adults.

The most suitable vegetation is a mown area of pasture, preferably with a shaded, sheltered sandy area where the chicks can scratch about and play to their hearts' content. The fencing should consist of tightly meshed wire, at least in the lower region of the fence, so that

The little chicks always enjoy an excursion into the open air. Even at the age of about two weeks, they stay very close to the mother and are happy to let her show them the best places to find feed.

For chicks of every age, whether newly hatched or already half grown, the mother hen is both a playground and a protector. How protected and delightful is the life of the brood shown here in the sun, the open air and the fresh greenery.

the chicks cannot leave the run, where they will easily fall victim to cats or free-running dogs.

If no access from the henhouse to the run is possible, it would be best to build or buy a mobile chicken coop which can be placed anywhere in sunny, feed-rich places. This method has the further advantage that the hygiene conditions in the run are much more favourable than the permanent pasture in front of a fixed henhouse, due to the frequent changes of pasture area that can be achieved. It does, of course, mean that the chicken keeper has to make a greater effort in looking after his birds and to take suitable care of them. The temperatures should roughly match the henhouse temperatures set according to the age of the chicks. In any event, good "pasture tours" are the best guarantee for successfully raising strong, healthy broods.

Helpful tip

It is a wise precaution to put wire netting over the entire run as a protection against enemies from the air. Wide-meshed wire can be used, or the plastic netting which is put over fruit trees in the autumn to protect them against marauding birds.

Critical life stages during natural and artificial rearing

The first few weeks of life are divided roughly into the

- *downy stage* = 1st to 3rd weeks of life,
- *feather-growing stage* = 3rd to 6th weeks of life, and
- *growth stage* = 6th to 8th weeks of life.

These periods can vary by up to a week, depending on the particular breed.

During the downy stage, the small birds need warmth above all, and a balanced diet, particularly containing protein. In our experience, toward the end of the first week is when the highest number of losses is to be expected.

The feather-growing stage is also marked by a relatively great need for warmth, whilst the requirement for space to roam, light, and fresh air should also be met. A healthy toughening-up process will make later action against illness largely unnecessary, saving both cost and stress for the chicken keeper. When feeding, as well as easily digested, protein-rich foods, sufficient phosphorus and calcium-containing feeds should be provided, in order to promote plumage formation.

The drive to imitate is very strong in small chicks. In this way, they quickly learn how to supply their own needs.

Finally, during the growth stage, care should be taken to provide an energy-rich balanced feed mix. The basic principle that applies here is that opportunities missed when the chickens are young are difficult to make up for in adulthood. Cutting corners in this regard is certainly not a good idea.

Rearing Adolescent Chickens

From the eighth week on, the young are no longer referred to as chicks, but as pullets and cockerels. Their plumage is now largely developed and is able to protect them adequately against the weather. We can now switch off the heating equipment. But care is called for because if the temperature suddenly falls, it is better to put the heaters on again and allow the birds a gentle transition to the coldest extremes. Usually the sexes are already separated during the 4th or 5th week of life – although this varies according to breed and are reared in separate flocks from then on. Although the cockerels, when they are kept in indoor conditions, are often fed a fattening-up diet and are destined for slaughter aged 10 to 12 weeks, the pullets are allowed to make extensive use of the run, so that they grow into robust, industrious layers.

Good to know

If pullets are used for breeding, between the 10th and 12th week of age is a suitable time for banding, since later on, the shanks become too thick.

Cockerels

The cockerels should, however, also be allowed access to a run, particularly if the keeper is not aiming to achieve rapid, intensive fattening of his livestock.

In any event, meat chickens raised in this way are not at all comparable with the cheap cockerels sold in supermarkets, which are fattened ready for slaughter in just five or six weeks. Rather, they are heavier, firmer fleshed and have a richer flavour.

Pullets

The female animals, to whom we naturally pay more attention due to their longer usefulness, pass through a phase in which they do not need so much care, although they do need more attentive management.

If there is not enough run space by the henhouse but there is sufficient room at some distance from it, it is worthwhile setting up a portable or relocatable young chicken house. Using this movable housing, which needs neither water nor a power supply, it is possible to provide fresh pasture at frequent intervals. This saves on feeding costs and offers the birds varied nutrition, increases their hardiness as they move about in fresh air and sunlight and, as a consequence of the frequent change of pasture, provides for better hygiene.

However, this method involves a greater time investment and a certain degree of risk. Water and feed must be brought daily to the site of the coop; after all, fresh, clean water is essential for healthy rearing, even when the birds are running free, and additional feed, which should be attuned to the pasture conditions (orchard, stubble field, harvested field), is highly recommended in this phase of rapid growth. But it is important not to pamper the girls excessively, otherwise they will neglect to seek feed for themselves. Diligent self-help can be trained in by specifically not encouraging a tendency toward laziness.

The risks of a free movement area of this type, whether spaciously fenced or without any fencing, have been the same since ancient times. The fox, with its tendency to live near humans, makes trouble as it always has, even in gardens on the edge of city and village alike. Hawks are less common, although they can be expected in some gardens. Much worse, though, are uncontrolled dogs and wandering cats, which quickly learn where the hunting is easy.

And account must be taken, of course, of neighbours who will not be exactly delighted over visits from the next-door hens scratching about in their herbaceous borders or defecating on their carefully tended lawns. Some good neighbourly relationships have come to a sorry end in court on account of the insufficient control of chickens.

Purchasing

From day-old chicks, pullets and fattening cockerels to laying-ready hens, we can always add to our stock by buying chickens. In any event, this issue always arises when one has just decided to become a chicken keeper. And in order to avoid unpleasant surprises, it is surely best to seek advice from an experienced neighbour or acquaintance. Local small-scale breeders should certainly be able to help. And additionally, as an important starting point, reference should be made to the "Judging Criteria for a Good Laying Hen" on page 12 and to the advice below from an earlier time, where in a book by the German author Robert Oettel from about 1873, it is stated:

One should satisfy oneself that the breastbone is not curved, this often originating in a hen's youth from the fact that the hens, before the bone has developed the necessary hardness, fall rapidly from a high perching rod. See, also, that the backbone is even and not hunched, that the hen does not carry the tail to one side, indicating a faulty development of the rump, and examine the tongue to ensure that it has not been shortened due to the method still practised in some places of tongue scraping to treat pip or pulling off the lower skin of the tongue, since every such reprehensible but often repeated operation always removes a piece of the tongue. Finally, one should check whether the claws are correctly present on the toes which, in the case of the cock in particular, is of great importance for self-evident reasons and also, whether the hen has a correctly-formed and not a crossed beak.

CHAPTER 8

Keeping Our Birds Healthy

The health of our animals should be particularly important to us. Sick hens lay no eggs, do not grow satisfactorily and are unsuitable for breeding.

Every chicken keeper should try to gain some knowledge of the most important diseases, their causes, their appearance and the possibilities for combating and healing them. Only in this way can he or she protect the animals against suffering.

Good to know

It is important to note that anyone keeping more than a certain total number of poultry on their premises must notify the statutory authorities in the UK – whilst for smaller numbers, notification is optional – so that the relevant agencies can take the necessary hygiene-relevant measures in the event of an outbreak of poultry diseases (such as avian influenza).

Since we can only deal in this book with the most important diseases, it is recommended to obtain a detailed book on poultry diseases. But even the knowledge obtained from specialist books does not mean that we will never need a veterinary surgeon. We would urgently warn against playing the role of vet in serious cases when symptoms first appear. Too much suffering has been visited upon unfortunate animals, although admittedly with good intentions, as the following example from earlier times illustrates:

Normally, however, the amateur poultry keeper pays little attention to his beloved pets when they are unwell. But if he spots something suspicious in an individual, then he thinks: "Aha, it has the pip; we can do something about that right now!" He fetches his pocket knife and cuts away half of the poor animal's tongue, wraps butter or pig fat round the excised tongue tip and stuffs it into the chicken's mouth, saying "Eat, bird, or die!" Often people take a feather previously greased with oil or fat from the preen gland and push it through the nose, after which they leave the animal treated – or rather maltreated – in this manner to its fate. The hen then settles sadly in a corner, to become healthy or not while immersed in its suffering.

In any event, it is useful to provide a separate henhouse section for sick and injured animals, or the possibility for temporarily setting up such a facility. It is, after all, often necessary for a while to isolate animals that have been given veterinary treatment so as to promote the healing process or even make it at all possible, and also to protect the healthy animals against infectious diseases. Following successful recovery, we can usually re-integrate them back into the flock without great difficulty.

The Healthy Chicken and the Sick Chicken

Before we consider the appearance of diseases, their possible causes and their proper treatment, we need, naturally enough, to know what a healthy hen looks like.

Bright clear eyes, smooth, shiny plumage and a light red comb and wattles are the first signs of good health. If we can hardly notice the breathing, there is no discharge from the eyes, nose or beak, these are further positive indications.

If we examine the animal more closely, we should find the throat mucosa to be flesh coloured or light red, but not blood red, pale or yellow. Defecation is also an important indicator. It should occur regularly – that is, about six times per hour – and should have a cohesive, not too watery, consistency. Diarrhoea is often recognisable from the soiled feathers around the cloaca.

A small group of healthy young birds enjoying a siesta.

All in all, a healthy chicken is easy to recognise: it is alert, it pecks and scratches, carefully tends to its plumage and eats and drinks regularly with a healthy appetite; in brief, it has a happy appearance.

A sickly chicken separates itself from the rest of the flock and sits apathetically in a corner with hanging wings, plumped-up feathers and a withdrawn head. It becomes visibly weaker and has dull, scruffy-looking plumage.

But we should not let things go this far, and in the next section we will examine how this can be achieved.

Prevention Is Better than Cure

Cleanliness, the right daily care and a balanced feed are the best means of ensuring

Good to know

The first signs of sickness often manifest themselves in a lack of appetite or, conversely, in excessive appetite and thirst.

that our birds stay healthy. And for this reason, among the important daily tasks in the henhouse is, above all, the cleaning of feed and water containers. It is also advisable to remove the droppings from the droppings board every week if possible, and when doing this, we can check its consistency and colour to watch for signs of disease. At the same time, we can clear the landing and perching rods of dirt.

Once a week, we should also examine the nests and, if needed, put in fresh bedding. In the scratching area, it is advisable to ensure that the litter is always loose, dry and dust free. About three or four times per year, particularly between the autumn and the spring when the animals remain mainly in the henhouse and the risk of infection from the ingestion of faeces increases, it is advisable to renew the litter.

The separate sand- or dustbath, which consists of dry, fine-grained sand, some wood ash and chalk dust, as discussed in an earlier chapter, should have some insect powder added from time to time to counteract annoying and often very persistent parasites. If a serious infestation threatens to take hold, we can give the chickens a thorough dusting with a treatment recommended by the vet.

When the henhouse is newly occupied, the parts of the hut and the hut equipment that have been movably constructed in accordance with the design in this book should be taken apart, cleaned and scrubbed down with hot washing soda solution. The walls should also be washed in the same way and then, after drying, painted with lime wash to give them a smooth surface, so allowing mites and other tormentors of the flock no opportunity to hide there. To be really sure, one can add a good – effective but safe – disinfectant to the lime wash. This arduous but worthwhile work reveals also how usefully the hut and its contents have been planned and constructed.

The Most Important Diseases

As has already been emphasised, the treatment of diseases is and remains the domain of the veterinary surgeon. The following brief outline of the nature of each disease, its appearance, possible treatment and recommended preventative measures, is intended, above all, to facilitate early recognition for the chicken keeper and at the same time, to impart a certain degree of fundamental knowledge, to make conversations with the vet more meaningful. But it is not intended to inspire the keeper to take independent action.

Aspergillosis

Aspergillosis is a fungal infection. The infectious agent, poisonous spores of different fungus types, is breathed in or ingested by the chickens from infected bedding or mouldy feed. The spores enter the lungs and air sacs, penetrate the tissues and cause serious inflammations.

In adult animals, the disease is difficult to identify. The signs that may be noticeable are a whitish pale face colour and a watery surrounding to the eyeball. Also diarrhoea, lassitude, wasting and wheezing are all characterising features of this disease. The duration of the illness is from four to six weeks. Deaths can be expected, particularly among young animals.

Treatment of aspergillosis is difficult and actually hardly achievable. However, sometimes the immediate transfer of the animals to clean, dry bedding can help. The most important preventative measures are bright, dry henhouses, dry and clean bedding, sufficient draught-free ventilation and scrupulously clean feed and water.

Good to know

In every case before new occupation and also regularly at least once per year, thorough cleaning of the henhouse should be carried out. Although it is an arduous job, it does the animals and our consciences a great deal of good.

Good to know

Since unfortunately Newcastle Disease breaks out again and again even in small flocks, vaccination by a vet is highly recommended.

Good to know

Unlike a broken leg, the treatment of fractured wings has little prospect of success.

Newcastle Disease

If Newcastle Disease is suspected, then the local Animal and Plant Health Agency must be notified at once. Under some circumstances all the birds may have to be killed.

The causative agent is the Newcastle Disease virus and transmission of the virus takes place mainly through the faeces and via nasal and throat secretions. The incubation period is approximately three to five days.

Symptoms include lack of appetite, high fever and green, liquid faeces. The sick animals drink large amounts of water, breathe gaspingly and often lay shell-free eggs (at night). In a more chronic course of the disease, neck twisting – where the head is placed between the legs or laid on the back – and walking backwards may be observed.

Eye Inflammations

Eye inflammations often occur due to colds and, particularly, as a result of draughts. The eyes run and may exude mucus, the lids are reddened and usually significantly swollen.

Using boracic eye lotion (2%), the inflamed eyes can be carefully bathed. The treatment must be frequently repeated. In particularly resistant cases, a vet should be consulted without fail.

Bumblefoot

Sharp-edged perching rods which are too narrow or a hard, stony run can be causes, among others, of injuries to the plantar region of the feet. Bumblefoot abscesses arise when these injuries become inflamed and fester. The abscesses can sometimes be confused with gout swellings by lay people, and the treatment, i.e. the cutting open of the swellings, should be left to a vet.

Fractures

Where bone fractures occur, it should initially be considered whether the unfortunate animal should perhaps be put down to relieve it from its suffering. The healing of broken bones is only possible if the fracture does not lie within the flesh, but is in the shank. The leg should be carefully brought into the correct position and splinted with the aid of wooden rods and adhesive wound-dressing tape. Usually, the animals can soon get mobile again and after two or three weeks, the splints can be removed. During this time they must, of course, be kept separate from the rest of the flock, but preferably in visual contact with them.

If bone fractures occur frequently in a flock of chickens, a lack of calcium and vitamin D may be the cause.

Infectious Bronchitis

The pathogen for infectious bronchitis is a virus that is released both through the droppings and nasal secretions and is spread very rapidly in a henhouse by dust and a variety of intermediate hosts. The birds become sick one to six days after infection.

Typical signs in infected chicks are shortness of breath, wheezing and sneezing. In the advanced stage, the birds have nasal effusion, lose weight and their plumage becomes ragged looking. The younger the individuals are, the more often deaths occur. In pullets and laying hens, the disease is harmless and has a short duration and in these animals, also, it is breathing difficulties that predominate.

A good henhouse climate, adequate ventilation, and care when buying birds are the most important preventative measures. The disease can also be prevented by inoculation.

Cholera

Fowl cholera hardly ever occurs nowadays. It is caused by bacteria and its occurrence is favoured by unhygienic keeping conditions, runs that are heavily contaminated with droppings, cold coops, and vitamin and calcium deficiencies. It is, however, easily prevented by good hygiene. The incubation period is four to nine days.

Infected animals lack energy, have no appetite and produce watery, sometimes bloody, diarrhoea. During the chronic course of the disease, there is severe shortness of breath and swelling in the joints. In the acute course of the disease, the animals die within a few hours or days.

Intestinal Inflammation

Simple intestinal inflammations, i.e. those which do not occur as concomitant conditions with other diseases, are usually caused by incorrectly mixed or spoilt feed. If this is the case, the feed must be changed at once and the animals housed in a warm henhouse. Some finely granulated wood charcoal can be mixed into the feed and the chickens given camomile or peppermint tea to drink.

Ovary Diseases

Diseases of the ovaries can occur either as independent conditions or as a concomitant with other diseases (leukaemia, typhus or avian influenza).

Good to know

The treatment of bronchitis is not possible, although its course can be made milder by the administration of vitamins and antibiotics.

Good to know

In order to prevent the spread of infectious cholera, the relevant regulations of the national veterinary authorities must be carefully followed. Any attempts at a cure for this disease are bound, in every case, to fail.

They can be caused, for example, by external influences such as jolts, blows and severe shaking. Laying activity slows down or stops altogether if the whole ovary has become morbidly altered. Diseases of the ovary are not curable and, for this reason, affected animals should be slaughtered.

Inflammation of the Oviduct

Inflammations of the oviduct can arise from copulation, incorrect palpation (feeling) of the hen by humans or by peck wounds during the laying process. The pathogens are usually coli bacteria or Salmonella, though viral infections can also cause the inflammations. Treatment is not possible, but clean and pest-free nesting litter is the most important preventative measure.

Frostbite

Chickens can tolerate dry cold without difficulty. But during extremes of cold, particularly in cold and damp coops, body parts which are not protected by plumage and where dangerous damp is able to settle can suffer frostbite: the comb, wattles, ear patches, toes and feet. The frozen body parts first become cold and empty of blood and then turn bluish-red and hot, after which they can swell severely.

A possible remedy is frostbite ointment. For prevention, the non-feathered areas, which are consequently at risk, can be rubbed with oil or fat. But it is much better to take the appropriate precautions.

Avian Influenza (Bird Flu/Chicken Flu)

This is a highly infectious disease. If an outbreak is suspected, the relevant veterinary authority must be notified without delay. The causative agent is the type A influenza virus, the most dangerous subtype or strain being H5N1. The virus is highly pathogenic, it affects all types of birds and no treatment exists. Particularly at risk from avian influenza are waterfowl and they are often the cause of a spread to other poultry livestock, especially chickens and turkeys. The symptoms are often similar to those of Newcastle Disease. If an outbreak occurs, the whole flock must be culled and disposed of in a manner that prevents further spread. The incubation period is only a few days. The best protective measures are avoiding contact with wild birds by means of suitable fencing and careful monitoring and hygiene in the run and the henhouse.

Gout

Gout most often appears in older animals with an excessively protein-rich diet and a vitamin A deficiency. It is a metabolic disturbance in which too much uric acid is formed in the body and cannot be completely removed by excretion. A distinction can be made between joint (articular) gout and internal (visceral) gout. In articular gout, swellings develop in the toes and foot joints, usually filled with a yellowish white mass (not to be confused with bumblefoot abscesses). In the case of visceral gout, the internal organs become covered with a fine-grained white coating.

Since, as mentioned above, this is a disease mostly of older birds, the death of the animals

Good to know

Animals that have suffered from frostbite should not be brought into the warm immediately, otherwise the frozen sites become distended and cause severe pain. It is often sufficient to rub the frozen parts thoroughly with snow and to massage them.

should not be drawn out with prolonged treatment. Plenty of access to the run, a generous amount of greenery and adjustment of the feed to a lower-protein diet are the best protection against gout.

Fowlpox

The causative agent of this disease, which mainly occurs during damp weather in autumn and winter, is the avipoxvirus.

Infection is through contact with the secretions of the oral and nasal mucous membranes and through faeces. The disease can also be spread rapidly through a flock by blood-sucking insects such as mites and ticks.

In the main form, pocks, which are wart-like nodules, develop on the comb, wattles and in the nasal region, on the ear patches and sometimes on the rest of the body. The pocks range from pea to cherry sized, are dark brown and fall off by themselves after some time.

The mucosa form shows itself as swellings with yellow deposits in the throat and the beak (with the danger of suffocation). Treatment is only worthwhile for mildly infected animals.

Favus

Favus is a fungal disease that affects the comb, wattles and ear patches and is passed from animal to animal by contact. Initially, small, light-coloured patches appear on the comb and these gradually spread and increase in size, becoming scale like.

At the first signs of this disease, the affected animals should be isolated and treated immediately. The scales should be rubbed with oily ointment, then carefully removed and the diseased areas painted with iodine. The hen-house and its equipment should be thoroughly disinfected to treat and prevent the disease.

Good to know

As a preventative measure, a fowlpox inoculation can be carried out on pullets.

Scaly Leg

This very commonly occurring disease is caused by mites (scaly leg mites) which bury themselves in the skin of the legs. Over time, thick scabs form which cause the animals difficulty when walking. The whole flock can become infected very rapidly.

The scabs can be softened with soft soap or vegetable oil and then gently washed off with lukewarm water. Following this, the legs should be treated with scaly leg ointment.

Crop Blockages

There are two main types, and the seldom-occurring *sour crop* arises when animals are given indigestible feed that has begun to ferment. The crop appears to be absolutely full and feels soft and elastic. The animals refuse feed and suffer from shortness of breath. It is worth trying to remove the crop contents by massage. To do this, the bird should be held with the head pointing downwards, so that the bad-smelling crop contents can flow away through the beak.

Impacted crop occurs when the feed is difficult to digest (hay, chopped straw) or when foreign bodies (bones, stones, wood shavings) have been ingested and block the outlet of the crop.

The crop protrudes and feels hard. Firstly, an attempt should be made to empty the crop (see above). If these attempts are unsuccessful, the crop contents can only be removed by an operation to open the crop carried out by the vet.

Leukosis

Leukosis is a widespread viral disease and infection can take place via the egg during brooding or by contact with other animals. The most susceptible individuals are young birds at the ages of six to ten weeks.

The sick animals gradually waste away, and the comb and the wattles become pale or yellowish. The liver is severely enlarged and has whitish flecks, which can also be present on the spleen and the kidneys. No cure is possible.

Egg Binding (see pages 110, 111)

Marek's Disease (Poultry Paralysis)

Marek's Disease is a viral infection which attacks the brain and nerves. Hanging wings and an uncertain, limping gait with unnaturally bent legs are among the signs of this disease. Other characteristic signs are grey-green coloration of the irises and irregularly shaped pupil edges.

Contagion is through infected chicks or pullets, virus-laden droppings, bedding or equipment as well as through ticks and mites. The mortality rate among young birds is up to 60%. There is no treatment for Marek's Disease, although preventative inoculation of one-day-old chicks is possible.

Rickets

Rickets is a disease of vitamin deficiency from which mainly chicks, but also adult chickens, can suffer.

Weakness of the legs, thickened joints and bending of the joints, buckling of the toes and squatting on the hocks are the outward signs. If rickets appears in the flock, immediate treatment with vitamin D3 should be carried out. Correctly fed animals with sufficient exercise in fresh air and sunshine will normally not suffer from this disease.

Coccidiosis

Coccidiosis is one of the most widespread of the diseases of young birds. It manifests as inflammation of the small intestine and caeca. The causative agent is a single-celled parasite, the resting forms (oocysts) of which are highly resistant to both heat and cold. It is mainly chicks in the second to eighth week of life that are affected by this insidious disease. Damp litter is an ideal breeding ground for the oocysts, which can remain capable of infection over long periods.

Transfer takes place by pecking in the droppings of infected animals and through soiled feed and drinking water.

The signs and the course of the disease can vary greatly. Three days after infection, the following symptoms appear: the infected birds become visibly sick, let their wings hang down and hardly eat at all. Bloody diarrhoea is passed. With a rapid course of the disease, death can occur after four or five days.

Treatment of the infected animals is usually not economic and is also risky, since they can become chronic carriers. If the disease is recognised early, however, its spread can be prevented by suitable means. The housing, equipment, and run must be thoroughly cleaned and disinfected.

Rhinitis

Rhinitis is transmitted by bacteria. A precondition for this infectious disease is almost always a cold brought on by damp warm or cold air in the henhouse, in conjunction with a vitamin A shortage. Pus-laden effusion from the nose, head shaking, sneezing, breathing difficulties, wheezing and squeaking breathing sounds (which used to be identified as "pip") as well as swellings in the region of nose and eyes are the characteristics of this disease. The incubation period is one to five days.

Tuberculosis

Tuberculosis appears mainly in flocks with a disproportionate number of older chickens and in poor, unhygienic keeping conditions. The disease can be very protracted and is infectious not only to other animals (cattle, horses, pigs), but also to humans. The sick animals gradually waste away, the comb and wattles becoming pale. When dissected, yellow nodes are found in the liver and spleen, on the intestinal walls and in the bones.

The causative agent (tubercle bacillus), which is shed with the droppings, is extremely resilient, but it is unable to withstand direct and prolonged sunlight.

Good to know

Regular removal of droppings and renewal of the litter, repeated disinfection of the hut and equipment, correct feeding of the animals and care of the run are among the most important preventative measures against coccidiosis.

Good to know

Healing is not possible in cases of tuberculosis among poultry. In the event of a serious outbreak, the whole flock must be slaughtered. The hut, equipment and the run must be extremely thoroughly disinfected. If possible, the henhouse should even be left unoccupied for a year.

Pests

As a result of blood-sucking mites, a lack of blood and consequently a greater susceptibility to disease and general constitutional weakness can occur. During the day, the mites remain in the henhouse, for example hidden under the perching rods or in cracks in the walls; at night, they come out and attack the poor animals to suck their blood.

Bird lice, *fleas* and *blood-sucking lice* are all skin parasites. They hide in the plumage of the birds, pestering and distressing them greatly.

All pests should be combated without delay and thoroughly, since they cause irreversible damage and torment the animals unnecessarily. Regular, thorough cleaning and disinfection of the henhouse and all items of equipment are the most important precautionary measures. Perching rods should be painstakingly cleaned and disinfected from time to time. Birds should be powdered at intervals with insecticides and given the opportunity at all times to use dustbaths. Liberal amounts of wood ash, tobacco dust or, now and then, insect powder should be mixed into the dustbath.

Pullorum Disease or White Diarrhoea

The causative agents of this disease, which is characterised by an extremely high mortality, are Salmonella bacteria.

Infection takes place, firstly, via the incubating eggs of infected hens, and secondly, uptake of the microbe from infected feed, in droppings, etc. The incubation period is two to five days.

The wrong keeping conditions during rearing (too cold or too hot) can cause the disease to break out suddenly. The chicks are weak and let their wings hang down, show a greater than usual need for warmth and refuse feed. Their droppings are white to greenish. The down feathers stick together round the cloaca.

Injuries

If an animal is injured and bleeding, it must be isolated from the rest of the flock, otherwise the sight of bleeding wounds stimulates the other chickens to constant pecking, which can degenerate into cannibalism. The wound must be carefully cleaned and thoroughly disinfected. The individual should remain isolated until the wound is completely healed, but if possible, with visual contact to the flock.

Poisoning

Poisoning can arise, above all, through negligence and lack of attention. Feed material must be stored completely separately from poisons and pest-control products and must not come into contact with them.

Chickens must not be allowed to eat corncockles, ergot fungus, feeds with a high salt content or spoilt feedstuffs under any circumstances. Creosote and similar wood preservatives in their liquid state can lead to corrosive burns and can be highly hazardous breathed in as a vapour. Mineral fertilisers (potash, saltpetre, nitrogen) can lead to life-threatening corrosive burns in the throat, crop and proventriculus. Cereal-disease control agents can also have seriously toxic effects.

Good to know

Treatment of pullorum or "white diarrhoea" is ineffective since, as chronic carriers, infected animals remain a constant danger to healthy animals. The most important preventative measures are hygiene and the correct temperature during rearing.

Poisoning is recognisable in that the animals retch, vomit, have diarrhoea and cramps, stagger or walk uncertainly. Drowsiness and excessive sleeping are further signs. Treatment of the affected animals always depends on the type of poison ingested.

Worm Infestation

Damp litter and a neglected run soiled with droppings are ideal breeding conditions for parasitic worms, but even in healthier environmental conditions, it is advisable to worm the birds at regular intervals.

- *Tapeworm* can grow to 15 cm long and are made up of individual segments. They are transmitted by intermediate hosts, usually flies and other insects.
- *Roundworm* and *capillaria worm* can seriously impair the development of young chickens in particular. The animals become wasted, suffer from diarrhoea and their plumage becomes ragged looking.
- *Gapeworm* or *red worm* manifests itself in chickens as a lack of appetite and shortness of breath. These worms attach themselves by suction to the mucous membrane of the trachea and feed on blood.
- *Heterakis gallinarum* or *pinworm/threadworm* are parasitic worms that attach themselves to the intestinal mucous membrane.

The Henhouse Medical Cabinet

It is a good idea to make sure that we are properly equipped for the preventative and treatment measures set out above, as well as for any emergencies that might arise. It is therefore recommended that a medical cabinet should be put together in good time with supplies from the vet.

Among other things, it should include:

- a multivitamin preparation for treating and preventing vitamin deficiency diseases (e.g. rickets) and for general strengthening, particularly during the growth phase and following a period of illness,
- a worming product for regular worm treatment,
- a product for treating diarrhoea-causing illness,
- a disinfectant for treating skin injuries and skin diseases,
- a product for treating ectoparasites such as bird lice, fleas and mites,
- a wound and frostbite ointment, and
- boracic lotion (2%) for treating eye inflammations.

Catching and Identifying Birds

If we need to treat a bird, it should be removed from the flock without causing a great stir. We cannot always assume that each of our chickens is so tame that we can get hold of it by hand whenever we need to. After all, whenever a bird notices that we intend to do something unfamiliar with it and the first attempt to grab it fails, it will be wary from then on. To catch it, we would then have to chase it through the henhouse or the run and – quite apart from the ridiculous spectacle we make of ourselves – in the process, unsettle the whole flock. It is better to use tried and tested aids such as the *catching hook* or the *catching gate*. The catching hook is something we can always use if we need to catch a particular hen from the group. For this, we can enter the hut or the run in a quite natural manner, move inconspicuously toward the particular chicken, place the catching hook carefully round one shank from behind and

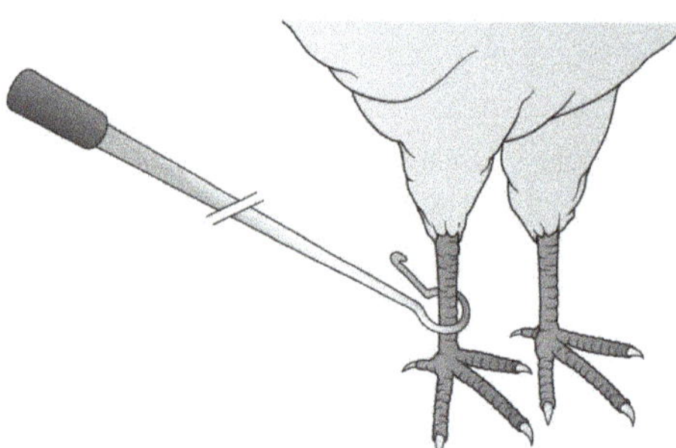

The catching hook makes it possible to pluck a bird from the flock quickly and reliably.

A catching gate works well for checking or where handling might be needed.

Good to know

Medicines should be stored shut away and out of the reach of children. Supplies and the use of pharmaceutical products should be noted in a henhouse stock book.

Numbered leg ring, pull-on type.

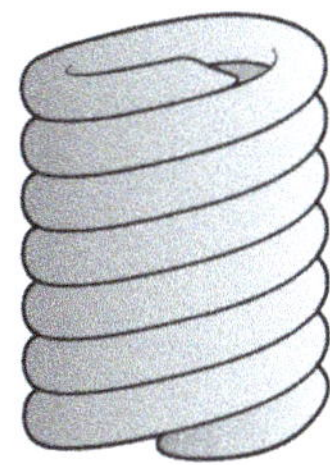

Coloured helical ring, put on by twisting the ring.

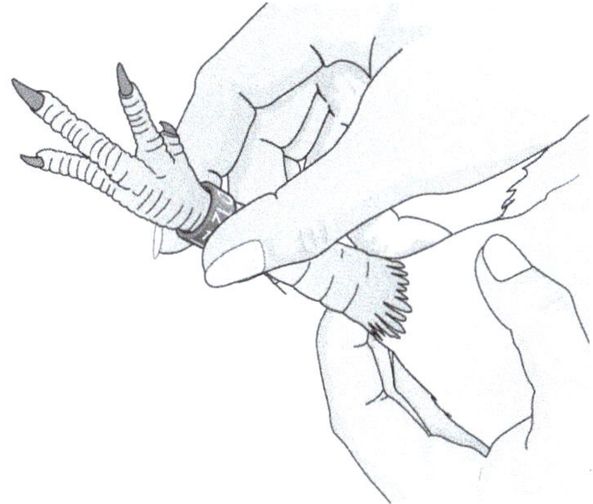

Leg rings simplify checking the output and the condition of our chicken flock.

then we have it in our power. At the same time, we should try, with a professional grasp of our free hand, to pull the chicken to us, so that it cannot beat unnecessarily with its wings and spread panic among the others.

The catching gate is worthwhile if we need, perhaps, to dust the whole flock with insect powder, carry out banding or wish to perform similar operations. For this, we drive the animals carefully into a corner using the opened catching gate and then place the two outer elements against the wall so that none of the chickens can escape. Now it is easy to take one animal after the other, treat them and put them back in the hut. In this way, we have the assurance also that no bird is treated twice or missed out.

This brings us to a further minor but important point: we cannot always tell our individual birds from one another without help. This is particularly the case during the moult time when their appearance temporarily changes completely. For this reason, the most useful and simplest solution is to use leg rings, which can be bought in a variety of forms. Owners of pure-bred poultry will naturally use numbered metal rings. For the hobby keeper without breeding ambitions, coloured helical rings made of plastic are quite sufficient. These have the advantage that they can also be applied later to adult animals.

CHAPTER 9

Poultry Produce

All About Eggs

The egg, the way in which it is produced and how it fulfils its purpose are truly wonders of the natural world and are well worth a closer look. For most people, the chicken egg – truly a mass-product of our day – appears commonplace and, quite frankly, banal. Of a broad variety, fresh, clean and inexpensive, it is readily available just about everywhere. But it is not likely that many of us stop to think about a "product" which, under certain circumstances, instead of finding its way straight into our kitchens, is able to bring new life into being. In all probability, most of us are not aware that by the time we eat it, if fertilised our breakfast egg would already have started to become a living being.

No one's enjoyment of eating cooked eggs should be spoilt by this realisation. It is intended merely to make us aware of the fact that what humans regard as an agricultural product is an original work of nature, regardless of whether it is the egg of a chicken in a battery cage, a deep-litter hen or a free-range bird: an egg remains an egg.

A Brief Cultural History

Though nowadays we regard the egg purely as a valuable foodstuff, in ancient times things were fundamentally different. Learned individuals like Pythagoras considered the eating of eggs to be an abhorrent act because they carry the seed of a new life in them. Other philosophers compared the egg with the four elements, so that the shell epitomised earth, the air cell represented air, the egg white water, and the yolk fire. Among many peoples, the egg was offered as a sacrifice at religious festivals, and that is where our painted or coloured Easter egg comes from. It is found in the ancient cultures of the Chinese, Egyptians and Persians, who named their new year festival in spring the "Festival of the Red Egg". Long into the Middle Ages, red, representing blood, love and also victory, was still

Some figures concerning chicken eggs	
Weight	58 g
Proportion made up by yolk	32%
Proportion made up by white	58%
Proportion made up by shell	10%
Shell thickness	0.2–0.4 mm
Breaking strength	2.5–4 kPa
Density of pores in shell	150/cm²
Freezing point	–2.2 to –2.8°C
Age of hen at start of laying	18–24 weeks
Laying output	225 eggs per year
Duration of passage through oviduct	22–25 hours
Per capita consumption in UK	184 eggs
Energy content per egg	85 kcal (= 350 kJ)

the most important and most authentic colour for Easter eggs. In Greece, Holy or Maundy Thursday is still known as "Red Thursday" because on this day the eggs for the Easter festival are mostly coloured red.

Others have used the egg as currency, as was the case in recent history, after the Second World War in Germany. Eggs were traded on the black market along with other highly desirable items such as cigarettes.

Famous chefs prize the egg as absolutely essential in the kitchen. It is used as food for the sick and plays a part in the success of many sauces, liqueurs, baked products and in mayonnaise. We wash our hair with it, use it for medicinal and technical purposes and, quite simply, eat it. The egg is even believed by some to enhance virility.

It would seem, then, that the egg is universally useful.

Formation of the Egg

How the egg is structured has already been described above in the section about the hatching egg. But the structure of an egg only really becomes clear when we consider how, from the tiny egg cell in the ovary of the hen, an actual chicken's egg is brought into being on its journey through the oviduct.

Origin

Recapping the description of the structure of the hatching egg, we can best begin this consideration of egg formation with the inner part of the object: the sphere of yolk.

This yellow ball, consisting of the germinal disk, the latebra, the yolk and surrounded by the vitelline or yolk membrane, is formed in the ovary of a sexually mature hen. The ovary of a healthy individual contains several thousand of these egg cells. They form, surrounded by a spherical cell layer, small yolk bubbles (follicles) which are attached by a stalk to the ovary, rather like a grape. When the yolk sphere reaches a certain size, the follicle membrane ruptures and releases it. If everything goes according to plan, the yellow sphere is caught by the funnel-shaped opening of the oviduct and the journey through the oviduct begins. This process repeats in laying hens every 24 to 36 hours and leads, in a further 24 hours, to the "birth" of a finished chicken egg. This makes it clear why each hen can only lay a maximum of one egg per day, a biological fact that is unaffected by either practical cultivation or by science. Writing about this in 1873, the German chicken-breeding pioneer Robert Oettel observed:

> *It may be possible to force plants, but just as one is not able to hatch an egg before the natural time, it is also beyond one's power to achieve 2 eggs in a day. As long*

From ovum to egg			
Oviduct section	Passage duration	Length in cm	Function
Infundibulum or funnel	20 minutes	8.0	Collection of the yolk-rich ovum and possible fertilisation
Magnum, main part of oviduct	2–3 hours	33.0	Secretion of egg white (45%)
Isthmus	1¼ hours	9.5	Secretion of egg white and formation of shell membrane
Shell gland (uterus)	20–21 hours	8.5	Secretion of egg white and formation of shell

years of observation have shown, each egg needs at least 20 to 22 hours to achieve full development and maturity and this is the very shortest period, even with the most fruitful breeds and only in exceptional cases, since with sustained daily laying, in most cases a hen lays 1 to 2 hours later each day. No hen is able to lay two fully formed eggs on the same day, even though it can occur that she produces a mature egg early and, in the afternoon or evening, a soft-shelled egg follows, this having as its cause over-excitement or some other irregularity; but the bird will then certainly not deliver an egg at all on the following day.

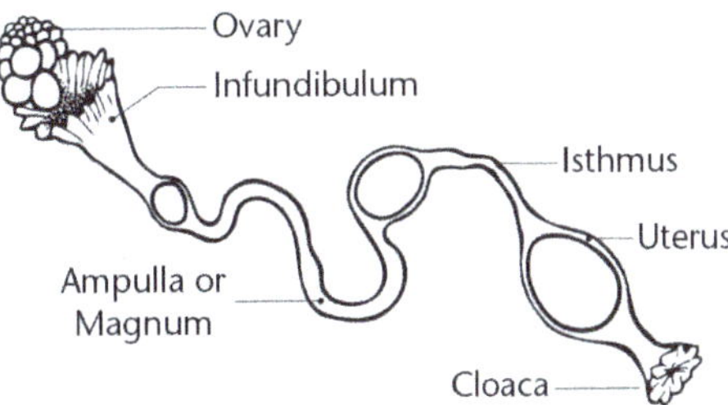

The egg has to complete this route through the oviduct before it lands, damp and shiny, in the nest.

Passage through the oviduct

During its passage through the oviduct, the ovum is supplied with everything that a hen's egg needs. If the hen has previously been covered by a cock, then the start of the journey in the funnel is also the place where the male sperm are able to fuse with the germ cell. Indeed, one mating is sufficient for several eggs, but that will be dealt with shortly. Soon the yolk sphere is surrounded by the first layer of egg white, specifically a viscous layer which forms the chalazae. The egg white is secreted by glands in the oviduct walls and, in the process, the yolk sphere rotates about its centre during the passage through the oviduct. During this process, the forming egg dwells for varying periods of time in the individual sections of the 60 to 75 cm-long oviduct and, progressing from the inside outward, gains the structure set out above in our description of the hatching egg.

The longest time, approximately 20 hours, is spent by the egg inside the shell gland or uterus. This is where it gains the finishing touch and is surrounded by its shell. Finger-shaped glands secrete a calcerous mass which surrounds the now finished structure, solidifies and becomes the enclosing shell. The cuticle is added and another work of wonder is complete.

Now the egg needs only to be moved to the outside of the body, and for this purpose it is pushed through the vagina into the cloaca, where it reaches the end of its journey.

The wall of the oviduct and of the cloaca turn inside out and a clean fresh egg lies damp and glistening in the nest, as announced by the hen with a loud clucking.

Abnormalities

Not every egg is so faultless as that described above. There is a variety of, sometimes curious, abnormalities that can arise:

- Firstly, there are the *soft-shelled* or *shell-less eggs*, known as *wind eggs*. This type has a soft shell or no shell at all and looks quite unappetising, although it is definitely still usable for human consumption, for example as scrambled egg or for baking. The cause may lie in a functional disturbance of the shell glands in the uterus or faulty feeding (insufficient calcium-containing supplement). Often this fault can be seen at the beginning of a hen's laying period.
- *Yolkless eggs* are formed entirely without a yolk. This condition can be caused by a nervous stimulation of the glands in the oviduct walls so that they produce egg white although no yolk sphere has passed into the oviduct, which in normal egg formation is the cause of this stimulus. Otherwise, the object that becomes surrounded with egg white is a matter of complete indifference to the oviduct. It is known, from experiments in which small cork balls of suitable size are implanted into the oviducts of hens, that quite normal eggs can be formed with, in place of the yolk ball, a ball of cork.
- *Double-yolked eggs* are formed when two follicles burst at the same time and their yolk balls slide together into the infundibulum of the oviduct. They are treated there like single-egg twins and are surrounded by the same egg white.
- *Contaminated eggs* contain foreign bodies which can enter the oviduct during the mating act, or blood spots which can become incorporated on the journey through the oviduct due to burst blood vessels or similar.
- Eggs can even be found which carry a finished egg inside them. This *egg within an egg* probably arises because the finished egg is retained due to physiological disruptions in the oviduct and makes a further partial passage through the oviduct.

All these variants are more or less edible curiosities. A version that is hazardous to the health of the hen, however, is the *ectopic yolk*, which can lead to egg yolk peritonitis. In this condition, the ripe yolk ball does not enter the funnel of the infundibulum, but passes into the abdominal cavity between the internal organs and, after a while, becomes putrefied. If several yolk balls have collected there, this leads firstly to the cessation of laying and finally to a painful death for the hen. The cause of this abnormality is usually the loss of firmness in the funnel. The reasons for this can be highly varied so that, for a lay person, it is better to cull animals with such disorders in their laying apparatus, rather than attempting what may well be a drawn-out search for the cause.

Egg Binding

Another problem that is caused by a malformation of the eggs is *egg binding*, which can result from excessively large, irregularly formed or sideways positioned eggs. Other causes may be a weakened or inflamed laying apparatus. This "difficult birth" often occurs during laying of the first egg.

Egg binding can be recognised by the fact that the hen makes bowing motions, lets her wings hang down and is clearly attempting, through desperate pushing, to relieve herself of her load.

In our own small flock, this tortured exertion certainly does not escape our notice. One can attempt, first of all – using an age-old method, holding the animal with its rear end over a steaming bowl of camomile water and by careful kneading – to massage the egg out of the oviduct. A douche with vegetable oil can also lead to successful results. It is certain that not everyone would wish to attempt performing such a tricky procedure on the unfortunate tormented bird. It is, in any event, better to leave the treatment to a vet to perform.

A Peculiarity of the Chicken

Finally, a word about the ovary of the hen. Whilst mammals typically have two ovaries in a paired arrangement, the chicken, like other birds, has only one ovary fully developed, specifically the left ovary. The stunted growth of the right organ can probably be explained by the fact that severe physical shocks in a bird's body, for example when landing, could cause eggs formed in two oviducts to collide with one another and shatter.

But Mother Nature has found a good remedy, although in this case not as she usually does, by the provision of apparent excess, but on the contrary, by cutting back on former excess.

Nutritional Value of the Egg

An egg is not a calorie bomb in the sense of being a highly energy-rich food, but it is a well-balanced combination of high-value nutrients coupled with good digestibility. It contains protein, carbohydrate, minerals, vitamins and fat, including lecithin.

The nutritional value of a single chicken egg of normal size is best illustrated by a comparison with the average daily food intake of an adult. What is noticeable is that the calorie requirement is only inadequately covered by it. But in its place, the content of valuable minerals and vitamins, high-quality protein and easily digested fats in a concentrated form as compared with other foods is unusually high.

In addition, the egg white contains less protein than the yolk. Interestingly, the white of the egg is traditionally the part associated with protein because of its conversion into a solid mass under the influence of heat, when the protein that is present in the white solidifies and, in the process, takes on a white coloration.

As far as human nutrition is concerned, there exist higher-value and lower-value forms of protein, made up as they are of amino acids. Among these amino acids are those which are known as essential amino acids. Essential means, in this case, nothing other than that they are amino acids that the human body cannot form itself but which are necessary for life and therefore need to be supplied from outside.

Composition of the egg	
Water	65.6%
Protein	12.1%
Fat	10.5%
Carbohydrate	0.9%
Minerals + vitamins	10.9%

Good to know

The egg is one of the most valuable food items. But considering that it has to serve as food for the journey, as it were, for a developing life, this is not really very surprising.

Composition of egg white and yolk		
	Egg white	Yolk
Water	87.9%	48.7%
Protein	10.6%	16.6%
Fat	–	32.6%
Carbohydrate	0.9%	1.0%
Minerals	0.6%	1.1%

Just as important and valuable for the harmonious progress of the life processes are the vitamins and minerals mentioned, with which the egg is also richly provided. Also not to be forgotten is lecithin, which equips the eater of plentiful eggs with a healthy nervous system, although it should not be expected to bring about miracles. Sensible incorporation of eggs into a balanced diet is advisable.

Egg Quality

A good egg is characterised by a variety of components. The most important of these are:

- appearance,
- freshness,
- colour,
- odour and taste, and
- size and weight.

Appearance

The appearance of an egg is essentially determined by the cleanliness, shape and quality of the shell. A clean egg has a more appetising appearance than a dirty or faeces-soiled example. Nowadays, however, it is sometimes the case that consumers are faced with a degree of charlatanism when partly soiled goods are advertised as organic or ecologically sound eggs, their only special connection with nature apparently being that they are naturally dirty.

It is also permitted – although for one's own consumption, it is of no importance – to clean dirty eggs before consuming them.

However, we should avoid washing eggs for too long or even soaking them in water. By doing so we might end up with a spotless egg, but for all that, they are then no longer clean on the inside because part of the dirt together with unpleasant microorganisms has now passed through the pores of the shell into the interior. It is therefore recommended to wipe clean dirty eggs clean with a cloth. If this does not succeed, the affected eggs may no longer be suitable for breakfast, but they are certainly usable for cooking and baking.

As far as the shape of eggs is concerned, the market demands oval, evenly shaped eggs with a smooth solid shell. This is important for mass production, especially due to the standardised packaging equipment and transport containers that are used. Moreover, with standardised shapes and smooth shells, fewer damaged or broken eggs can also be expected. The small-scale chicken keeper and grow-your-own enthusiast does not need to adhere to such strict standards, however. He can rejoice over each egg as it is, and can even tell from the shape and exterior quality which hen laid it.

Freshness

The degree of freshness of an egg is highly regarded by the consumer as a mark of quality. The perception of freshness is directly associated with a particularly good taste or an especially high healthiness value, the underlying principle being: the fresher the better. We know that when they are suitably stored, eggs will keep for quite some time without any great loss in the constituents that give them their value. However, this shelf life has its limits, which, given the present-day status

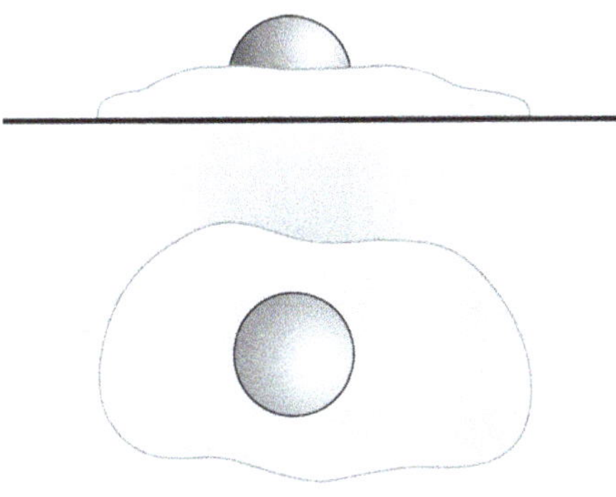

Fresh egg. The yolk lies almost in the centre of the white.

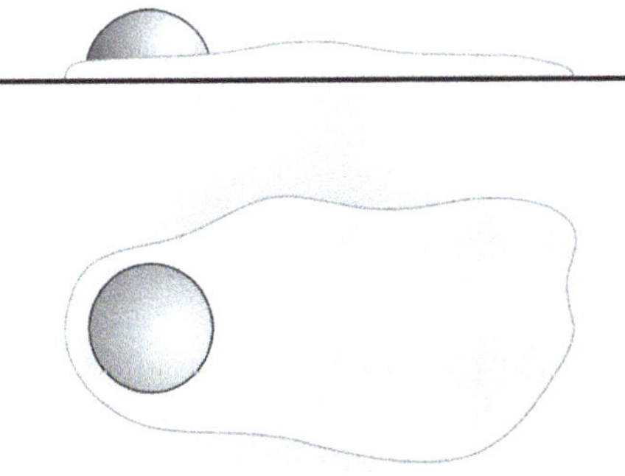

One to two weeks old. The yolk runs to the edge of the white.

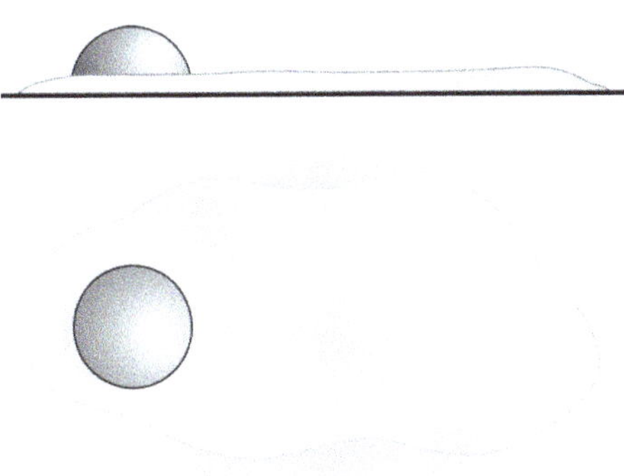

More than two weeks old. The white has a watery consistency.

Good to know

Eggs from chickens with access to a run or kept free-range can have a clean outer appearance if the supplier keeps his chickens in an appropriate way.

of egg supply, do not necessarily need to be exhausted.

The reader of this book who is either already a small-scale chicken keeper or wishes to become one, might well object: "What's that to me? I will always have fresh eggs from my hens." In principle, he or she would be right, but at particular times – either during a long absence, during the moult, or in winter – he will surely find it necessary to store eggs and to keep them as fresh as possible. But if he wishes to sell or give away excess eggs, the following suggestions will be useful.

The simplest method for testing eggs for age, and therefore freshness, is to place them in a pot filled with salt water and watch how they behave. If they remain lying flat at the bottom of the pot, they are only a few days old, whereas at the age of about a week, they lie with the blunt end pointing obliquely upward and at one to three weeks old, they will lie vertically in the water with the tip downward. Eggs that are yet older will float and should only be consumed with caution.

The reason for this age-related behaviour is the varying size of the air cell, which is usually situated, as we know, at the blunt end of an egg. It grows larger as the egg becomes older, since the egg evaporates liquid through the pores and this liquid is replaced by air. It follows from this that the method described can only provide an approximate guide for the age and freshness of any given egg. After all, it gives off moisture more or less rapidly depending on the storage conditions.

Characteristics of fresh and old chicken eggs		
	Fresh	Old
Shell	shiny	dull, lustreless
Air cell	small	enlarged
Egg white	mainly viscous fluid	mainly watery fluid
Yolk	lies in the centre, hardly mobile, holds together, high, skin taut, uniform colour	movable, sometimes also sticks to shell, flat, flows easily, skin folded or creased, cloudy appearance
Odour	none	musty, later unpleasant-smelling
Taste	aromatic	bland, increasingly foul

Even when we crack open an egg to prepare a delicious fried egg, we can tell at one glance whether what we are looking at is a fresh product or is already old. A fresh egg has its yolk steady in the middle of the white and is held together by it in a concentrated space. Furthermore, the different layers of white are clearly recognisable. After a week, the whole thing looks less stable, the eye of the yolk begins to wander to the border of the white and the layer of white is flatter. After more than three weeks, what we have before us is a rather suspicious-looking watery eye that makes a somewhat sad sight in the pan. It is still edible, without doubt, but given the appearance of such an old egg, the enjoyment of it is certainly not undiminished.

Colour

Two components can be distinguished within this quality indicator, specifically the colour of the shell and that of the yolk.

The colour of the exterior of a chicken egg surely cannot be a serious objective criterion for nutritional quality, even though eggs with a brownish shell coloration always enjoy greater popularity because, in the minds of consumers, they are associated with ecologically sound or organically produced eggs, on account of their supposedly natural colour. This prejudice cannot easily be changed.

In general, what is desired is an egg with a yolk that has a saturated golden colour. The substances responsible for this are carotenoids, natural colouring agents that have a yellow, orange or red hue and are contained to a greater or lesser extent in the birds' feed. Depending on the combination of these colour-imparting components, a relatively light or dark yellow colouration is produced.

Here again, we can refute another prejudice. The yolk colour is no indication of keeping style, although the attempt is often made to deceive consumers about this. Cage-free chickens with access to a run are certainly capable of laying eggs with only a pale yellow yolk colour. Of most importance for achieving a golden yolk colour in this situation, is the provision of feed, again, with the above-mentioned colour-imparting components, which means having a green area in the run with a high proportion of carotene-containing plants or a suitable basic feed with a high proportion of green lucerne meal or red pepper meal.

In many industrially produced mixed feeds, in place of the natural carotene sources, artificial colouring agents are used. They have the advantage that their colouring effect does not fade even after long storage. In both cases, the effect of the feed mixture on the yolk colour is astounding.

It is even possible, using suitable colouring agents, rather than yellow yolks, to create yolks that are green, blue, red or other colours.

But for us, as keepers of a small flock, the colouring substances provided by nature in the greenery of the run are certainly sufficient to provide a pleasantly yellow yolk colour on the plate or in the egg cup. We can, if we wish, help our hens in the arduous task of yolk colouring with a suitable proportion of maize meal in the feed and, particularly in the winter, by supplementing chopped carrots, which are well known to contain the desired colouring agent in a high concentration.

Good to know

The colour of the exterior of a hen's egg is purely breed related. Thus when selecting the breed we wish to acquire, we should always take the egg colour into consideration as well, if this aspect seems significant.

Odour and Taste

Just as with wine, in the case of eggs also, these two aspects of quality should not be regarded separately from one another. Whilst with the former, there is a broad spectrum of objectified subjective impressions – as everyone who has watched food and drink programmes on TV or participated in a wine tasting knows – things have not developed to that extent in the appreciation of eggs. Unless, that is, anyone has ever heard of a "dry Leghorn", a "fruity Sussex" or a "refreshing Rhinelander". Even if some assert that eggs from free-range chickens taste better than battery chicken eggs, no one has yet been able to show proof, for lack of the criteria for judging taste in relation to eggs.

Eggs clearly react sensitively to substances with a strong odour when they come into contact with, or are stored close to, them. Due to the numerous pores in the shell, which are intended to ensure gas exchange, they readily absorb strong aromas present in the air. Thus, for example, an oily fish flavour may be caused by eggs being stored close to feedstuffs containing fish meal. An unpleasant henhouse smell can also transfer itself to the eggs, and for this reason, eggs should be collected as soon as possible after laying and then stored in a well-ventilated, odour-neutral location. In addition, it is advisable to avoid feed mixtures that have a noticeable odour of their own. By following these simple ground rules, we can make sure that the eggs will always benefit from an untainted flavour.

Size and Weight

These quality criteria are naturally also inseparably linked to one another. If we wonder why our hens lay not only differently shaped, but also differently sized eggs, there is a variety of possible answers to this question.

The first criterion is certainly the breed. Large hens generally lay larger eggs than smaller, lighter or miniature breeds.

The size of an egg is also dependent on the age of the hen, young hens laying significantly smaller eggs than their older coop-mates. And within a particular age-group of the same

Good to know

The odour and taste of a chicken egg depend mainly on the feed given to the hen and the storage conditions of the egg.

Average egg weights and constituents in comparison				
	Weight in g	Proportion made up by yolk, %	Proportion made up by white, %	Proportion made up by shell, %
Goose	161.0	35	52	13
Turkey	86.0	33	56	11
Duck	70.5	36	54	10
Chicken	58.0	32	58	10
Pigeon	19.5	19	71	10

breed, there can be significant differences since the physiological capabilities of an individual also differ. We have noticed this again and again among our own hens. Whilst egg size and therefore also weight can only give a quantitative measure for the quality of an egg, it is certainly interesting to know that as the size of an egg increases, its percentage composition varies also; the relative proportion of the nutritionally important yolk becomes smaller to the same extent that the proportion of egg white increases. With a weight difference of 20 g, the proportional shift is approximately 4% by volume. This is something that the consumer is often not aware of and is essentially not important for him or her because, in a larger egg, the absolute weight or volume of the yolk is just as great or greater. Only the shell retains the same relative proportion in large and small eggs. Here also, the optimum in the price–performance ratio lies in a happy mean.

Quality Grades

For this reason, egg prices are not based on the percentage proportion of the value-creating constituents of the content, which is difficult to evaluate, but rather on weight. In this regard, binding standards have been laid down for the European Union area which no longer differentiate according to weight classes 1–7, but rather, based on EC Regulation No. 1511/96 of 1 August 1996, according to the familiar sizing system for clothing using the categories XL, L, M and S. This new weight or size system is defined as set out in the table below.

Apart from the usual obligatory details regarding egg packaging, further information may optionally be printed on, such as the keeping style or the laying date, so not only the "best before" date (A-class eggs) or the packaging date (B-class eggs).

With regard to the production style, only the following designations may be used:

- 0 = organic eggs,
- 1 = free-range eggs,
- 2 = deep-litter or barn eggs, and
- 3 = battery-cage eggs.

These keeping styles are also precisely defined in an EC Regulation. And therefore, other designations such as "meadow eggs"

Egg size and weight categories according to EC Regulation		
EC classification	Size description	Weight class
XL	Extra large	73 g and over
L	Large	63 g and up to 73 g
M	Medium	53 g and up to 63 g
S	Small	under 53 g

or "animal welfare compliant production" are not permitted.

Consumers tend to prefer the medium-weight classes with a tendency toward the larger end, since this size is closest to "egg cup-sized" and seems to offer the most balanced price–performance ratio.

There is a further differentiation according to freshness and prior handling into C-, B- and A-class eggs, with eggs of class C being intended for industrial processing, those of class B having already been stored or made storable for longer, whilst those of class A corresponding to the modern-day standard for a fresh breakfast egg. If we were to consider selling the eggs from our flock directly commercially, we would have to observe the currently applicable marketing standards for eggs, which regulate in detail the handling of the eggs from identification and storage through to transportation.

The bottom line for us is to be happy with every egg of normal size and condition that we receive and we have no wish to possess hens that lay the largest possible eggs. This would not be not useful for us as self-suppliers, nor would it offer us, as small-scale producers, great sales opportunities. And if the eggs from our flock become ever larger, then that is merely a sign that the hens are gradually reaching the end of their egg-laying time and that the flock urgently needs to be rejuvenated.

Consequently, the medium level of quantity is where higher quality and economic effectiveness can be achieved.

Consumption Possibilities

Countless consumers of eggs have surely dreamed all their life of eggs fresh from the nest, although it may be a small consolation for those who do cherish this dream that, according to the results of scientific investigations and the authors' experience, really freshly laid eggs have not yet reached their optimum flavour.

This is also a pleasure that the average consumer is hardly likely to experience since, until the eggs reach him, they are certainly already more than a week old. The small-scale chicken keeper, however, can enjoy this luxury day after day. But he will probably not be able to use all the eggs his hens produce at their flavour optimum, and will need to store or preserve them suitably.

Although the storage methods of earlier times hardly have any relevance for us given the plentiful daily supply of chicken's eggs year round, it is still important to know how such a sensitive protein-rich structure can be kept for long periods.

People used to use earthenware pots filled with a liquid preserving agent such as waterglass, salt water or limewater. Completely clean and undamaged eggs were laid in the liquid and were then kept "fresh" and consumable over several months in a suitably cool but frost-free environment. Depending on the preservation method and the preservation time, compromises had to be made with regard to use in the kitchen. Eggs preserved in salt water, for example, are no longer as suitable as fresh boiled eggs for breakfast; they would also not be used for making whisked egg white in gourmet cookery. A means of preserving lay in sealing the pores of the eggs with a variety of materials. For this purpose, among other things, people used fat, paraffin, varnish, glycerine, salicylic acid solution, glue, gum, shellac and flavourless Vaseline.

It is clear that people stopped at almost nothing in times when eggs were scarce, even with the readily imaginable sacrifice in the flavour. The fact that the methods mentioned were subject to some criticism and that

Good to know

The optimum time for the greatest enjoyment of a hen's egg is between the third and fourth day after laying, assuming it has been kept in optimum storage conditions.

everyone knew better is demonstrated by the following passage:

A great deal has been written and argued over the most suitable manner for preserving eggs as long as possible for commercial use. It is a prejudice to believe that eggs laid during the dog days of summer keep longer; naturally, the most recently laid eggs always keep the best of all. Eggs can be kept for several months, provided they have not been incubated and do not have any cracks, on an ordinary egg rack, whereas it is not to be recommended that they be placed on straw because they easily take on a musty overtone; they must be kept in a cool, dry place. Placing them in boxes on wheat bran or sieved wood ash so that they do not touch each other is another suitable method of storage. But the surest method remains limewater, which is used with great success by persons who involve themselves in the trading of eggs. This is prepared by boiling water in order to drive off the carbonic acid and atmospheric air therefrom and then by dissolving freshly fired lime in it, stirring the mixture several times, and then pouring the water onto the eggs situated in pots, so that a coating of carbonic lime forms round the shell, which prevents the ingress of air. Not only does this limewater need to cover the eggs entirely, but should even lie several inches deep above them. The eggs to be preserved in this way must not only be free of dirt, but there should also be none among them that are damaged, since they would first turn foul and then cause the spoiling of the others.

Meat

Meat from chickens is wholesome, nutritionally valuable and tasty. And therefore, alongside establishments for egg production, special meat production facilities have developed where, starting with one-day-old chicks, within six weeks, market-ready meat poultry are produced. Commercial meat production involves only young animals that comply with the demands of consumers for low-fat, protein-rich chicken meat.

Fattening and Meat from Young Birds

As stated on p. 98 in the context of incubation, it is a fact that on hatching, about half of all hatched chicks are male and half are female. The question of what to do with the cockerel chicks therefore arises.

It is surely most economical from our standpoint to fatten them ourselves, and, in this way, aside from the soup chickens that we find we have from time to time, to have some tasty roasts. If we want to raise offspring ourselves but also wish to have some good meat-producing birds, it is worthwhile buying young cockerels from a breeder and fattening them up.

In this case, we could proceed in the same way as the commercial meat producers and utilise the rapid growth phase of the young birds by feeding them with concentrated fattening feed and keeping them in a restricted space to obtain a commercial-type cockerel in the shortest time possible. We can also give ourselves more time by allowing the young animals to grow somewhat older and larger and only subjecting them to actual fattening in the stage between their 12th and 14th weeks. The pre-fattening phase in the first 12 weeks corresponds to the usual rearing time of young chickens, with corresponding run access or at least adequate opportunities for moving about in a well-lit, dry henhouse.

For the final fattening in the last two weeks, the feed rations set out in the table below, by way of example, should be given, and during this time, the birds should be kept exclusively in the henhouse.

Daily fattening ration for a chicken of a medium weight type	
Cooked kitchen waste – particularly potatoes	60 g
Wheat and oat bran	20–30 g
Barley grains and chopped maize	as wanted

The very sight of such a juicy piece of meat arouses anticipation of the pleasure of eating it.

The result is impressive, namely meat of a firm consistency, a low water content and in a generous quantity. The birds we fatten up ourselves produce meals to satisfy a small family, in contrast to typical supermarket roasting cockerels, which are sufficient only for two people at most. From experience, we can say with assurance that it is a joy to prepare a young bird carefully fattened by this method for cooking, and a genuine pleasure to eat it.

In general, however, it is important – and this applies for the fattening of older animals also – to keep a breed of chickens that, as well as having good egg-laying performance, is also a good supplier of meat; in short, a real all-rounder.

It is also advantageous, although this depends on personal taste, if the bird has the lightest possible meat and light-coloured skin, which in our part of the world is regarded as more appetising than yellow or dark flesh with a corresponding skin colour.

Fattening and the Meat of Older Birds

Commercial fattening of fully grown hens and cocks is not a significant business these days, although things were different before the mid-20th century. Birds for fattening were always seen as those that had completed their growing phase and had a correspondingly well-developed padding of fat. At that time, the chicken did not serve so much as a supplier of the leanest possible protein-rich meat, but also as a relatively economical source of fat for cooking. It was as much a part of the cooking and culinary culture of that time to fatten fully grown chickens, just as today it is to feed up young birds for the table.

No effort was spared in pursuit of achieving a full-fleshed and fat-padded bird for slaughter. Indeed, people competed with each other in contriving new, or refining old fattening methods, and this sometimes escalated to various brutal practices. Whilst the generally acknowledged method during final fattening was force-feeding, some crazed individuals went as far as to tie or even to nail the birds to their perches for fattening so that they would not lose a single gramme of their gained weight through unnecessary movement, or they dug out the poor creatures' eyes with the intention of reducing their urge to move about, or kept them for the same reason in solitary confinement in cramped cages in dark, unventilated corners.

Apart from these perverse excesses, force-feeding – as mentioned above – was the most commonly practised method of fattening poultry. For the economically minded poultry keeper aiming to supply consumer demand for fattened poultry in the form desired at that time, this method was merely the logical solution to a problem. After all, beyond a particular time point, the chickens did not, and do not, want to eat more than is good for them. And this was exactly the time at which force-feeding was started for the full fattening phase. What this actually looked like is described by the following original text:

> *The force-feeder takes two birds at once, wraps them in a cloth leaving just their heads free and places them on her lap. Beside her is a vessel with water and a bowl with the dough. This dough is thick; from it are made balls the size of an olive (or a large cherry) and these are forced down the birds' throats until their crops are full.*
>
> *To facilitate the forcing, one bird or the other is taken alternately. When one has swallowed two or three lumps, the other bird is then taken, and so on. The lumps are first dipped in water and then pushed into the beak, making their descent easier. Only rarely is a chicken given more than 15 lumps at one meal. Finally, the lumps have to be pushed down with the thumb so that they reach the crop. The dough is made from coarse maize meal and fresh milk, though alternatively, a meal made of buckwheat, barley or beech nuts can be used.*

Now the reader of this book might be asking him or herself what all of this has to do with our modern, relatively humane, fattening methods and particularly with the probably more elaborate keeping system of the small chicken keeper.

The answer to this question is as follows: firstly, it is probably interesting, and perhaps useful, for present-day chicken keepers and consumers to know how people in past times, to whom we tend to ascribe a closer connection to nature, sometimes acted with great cruelty toward living creatures for their own advantage. Secondly, it should remind us that although we can make the route to our actual aim of obtaining tasty meat to eat more bearable or even pleasant for the animals themselves, we cannot do away with the bitter conclusion – their violent death. And thirdly, we should consider that the meat of chickens in particular represents, for many of the world's poorer people especially in developing countries, a highly valuable and usually also affordable source of nutrition. The latter is an aspect that is often given too little consideration in the discussion about humane methods for the farming of animals.

A Practical Approach

But now to return to our wish for and the possibilities for obtaining from a modest flock of chickens, in addition to some delicious eggs, a succulent roast or two. As Wilhelm Busch put it:

Not just with the eggs they lay us,
for the care we take repay us,
but also since just now and then,
we can dine on roasted hen.

As a rule, we keep chickens mainly for "the eggs they lay us" and regard the occasional "roasted hen" as an added treat. But paying regard to the above, it would be a waste and also not clever to fail to utilise such a healthy and tasty source of nutrition to the full.

With a little planning, it is certainly possible to gain a respectable amount of chicken meat from a small flock.

If we assume that our annually produced offspring will be usefully incorporated into the flock – that is, we will have one- and two-year-old chickens mixed together in a 50:50 ratio – then from time to time, we will have individual animals of both age groups which, due to poor laying performance, unacceptably aggressive behaviour or injuries sustained, have to be slaughtered. It should be mentioned here that these are not birds that are regarded as surplus offspring and are raised as young meat-producing animals. One-year-old hens are certainly good enough as broilers following a suitable short fattening period and, depending on breed, can certainly provide a satisfying and tasty meal. If any animals from the two-year-old group have to be retired, it is better to plan in advance to use them for soups or stocks or to make tender stews from them.

Good to know

Of the cereal types, barley as whole grains and maize, broken up if at all possible, are particularly effective for fattening.

Fattening Feed

For one- and two-year-old birds, just as for young meat birds, it is best to use a feed with a high proportion of carbohydrates both for rearing and also for egg production from the laying hens. The proportion of starch and fat-containing cereal types should also be increased at the expense of protein-rich components (for example, soya meal, pea meal, etc.). If we wish to make things easy for ourselves, we can use a ready-mixed fattening feed from a commercial feed source. In our own case, where there are birds to be fattened only at particular seasons (fattened young birds) or in individual cases (due to culling and other causes), it is certainly more suitable to create our own fattening mixture.

For this purpose, kitchen waste mixed with cereal meal and bran can be used. Cooked potatoes, in particular, are enthusiastically consumed by the chickens. It is important, though, not to give them too much potato, since due to its high water content, compared with the same quantity of cereal, it contains significantly less energy. As a substitute, dried potato flakes can be bought; these exceed the energy (carbohydrate) content even of cereals. Unfortunately, potato flakes are not cheap and it is therefore only sensible to use them in high output mixtures in the form of dry feed.

Cooked potatoes, carrots and other root vegetables occur frequently as waste in a healthy kitchen and are therefore ideal as supplementary feed items in our little fattening establishments.

The Timing of Fattening

Now the question arises as to the best way to supply this special feed mixture to the birds selected for fattening and from which keeping style they – but mostly we – will profit most. Firstly, we should consider that we will never have large numbers of chickens that we can put onto a fattening diet. After all, our own henhouse space is limited to 12 hens and one cock. And this means that, at least during the winter when we usually have the smallest number of birds, it is only in rare instances that we have birds we would consider fattening. It would therefore hardly be worthwhile to use an individual case, arising through a drop-off in laying performance or due to increased aggressiveness, as the occasion to start a short fattening regime. Added to this is the fact that during the cold months, the energy requirement for bodily maintenance is greater and therefore fattening becomes more expensive.

It is mainly in the late autumn and at the beginning of winter, following the fattening of the young birds – if we are doing any fattening – that a further brief fattening period for any animals weeded out on account of poor laying or developing bad habits would suggest itself. In these cases, we are grateful for the benefits of a freezer.

Good to know

When choosing the right time for fattening, it should be ensured if possible that the animals are not in the process of moulting, otherwise plucking becomes an ordeal.

Keeping Birds for Fattening

The preconditions for productive fattening are – as mentioned above – a suitably matched feed mixture, calm surroundings and as little movement as possible. The feed composition has already been discussed and need not be considered again. In order to achieve sufficient rest and restricted freedom of movement, we should remove the hens intended for fattening out of the flock and, if possible, put them in another hut or at least in a separate henhouse compartment. In this setting, given light and air and unmolested by the hurly-burly of the henhouse, the birds should rest and feed, feed and rest, until they put on a good amount of weight in the form of meat and fat.

But it doesn't end there. If we are to believe the experts of former times, the flavour of the meat can be influenced by feeding various seasonings even before it reaches the roasting dish. We have never tried this out ourselves, and there is no need to stuff the poor creatures, but ingredients can be added to the feed in a ground or crushed form. One expert on the subject wrote the following:

> *Finally, a week before the end of the fattening period, one must consider the aromatising of the bird, according to the preferences for consumption or the taste of the buyer. For this process, the force-feeder should make dough balls of barley meal or bran and a meal of the following substances: cinnamon, coriander, juniper berries – whole or crushed – kneaded to a dough with a little barley meal and milk. With practice and experience, one learns*

the necessary quantity of the aromatic ingredients to produce an aroma which is all the more delightful for being barely noticeable.

Slaughtering

And so we come to a topic which is not exactly a favourite among many small flock keepers: the slaughtering of chickens. It is a deed that needs to be done as painlessly as is humanly possible. It is a fundamental principle that the animal must be stunned beforehand. This measure is also prescribed by law (Welfare of Animals Regulations in Great Britain) and must be complied with. For at least half a day before slaughtering, the animals must not be fed, so that the crop and intestines contain as little feed as possible.

Before doing the deed, we should have prepared a wooden chopping block and the slaughtering utensils, such as a sharp hatchet, and also a sharp knife on a small table, a bowl containing warm water, an empty bowl, a clean tea towel and next to the chopping block, a bucket. All this equipment should be brought together at a location that is not freely accessible – particularly for inquisitive children – and is positioned apart from the other chickens, so as to spare them any agitation.

Stunning is best performed by means of a well-aimed powerful blow with a round wooden baton to the back of the bird's head.

Composition of the slaughtered chicken	
Live weight	100%
Blood and feathers	13%
Head, feed and internal organs	17%
Edible organs	6%
Meat	52%
Bone	12%

To do this, the chicken is grasped by both shanks and held with the head downwards until it becomes quiet and bends its neck a little backwards and upwards, so that the blow can be delivered in a targeted manner.

Immediately after this, the actual killing can take place by severing the head from the body using a sharp hatchet on the chopping block. Directly after this, the chicken should be held in the bucket until it is bled dry, so that the person doing the slaughtering can avoid being spattered with the blood as it flows out.

Plucking

Once the blood has flowed out, plucking can begin while the chicken is still warm.

It is important to note that the more easily the feathers can be removed from the still-warm bird, the cleaner and more appetising the skin will be after preparation. Even if there are several birds to be slaughtered on the same day, this rule should always be followed.

If, however, the work is interrupted and the slaughtered bird becomes cold, a hot water bath can be used to help by immersing the chicken and allowing the hot water to have its effect. Scalding with boiling water has a more intensive effect, but also has the disadvantage that the skin becomes fragile and during subsequent plucking, easily tears, making the end result less appetising.

Plucking is carried out either standing with the bird lying on the table or sitting with it on one's lap, using the right hand while the left hand holds the shanks steady, in the following sequence: first of all, the flight feathers of the two wings are plucked out, followed by the tail and then the breast feathers working toward the tail, the back moving toward the head, and finally the thighs, neck and all remaining parts. Any remaining "hairs" or quills can be taken out by grasping them between the back of a knife and the thumb and yanking them out with a jerking motion. Finally, the naked bird can be singed over a spirit flame to remove any little bits of the plumage that still remain.

Evisceration

For evisceration, it is best to begin with an incision at the base of the neck, which should be continued as far as the throat. Using poultry shears or the sharp knife, separate the spinal column at the lower end of the neck. The index finger of the right hand can now be inserted and the internal organs loosened by moving the finger around inside the carcase. A cut should then be made between the cloaca and the tail. Continue this incision carefully around the cloaca. It is important to make sure that the rectum is not damaged during this process.

If the cut around the cloaca has been successful, then the intestines attached to it can be pulled out. The gizzard and proventriculus follow, along with the previously loosened viscera such as the heart, lungs and liver. Care should be taken with the liver because it has the inconspicuous gall bladder attached to it and this is easily damaged, releasing the unappetising and clearly visible green gall. This must be removed carefully. An error at this stage, spilling the gall onto parts of the innards or even the meat, could make them inedible. The carcase should now be rinsed under running water and is then ready for preparation.

A Few Recipes

And now to the real pleasure. The reward for all the sweat and toil should, after all, be pleasure for the palate and a hungry belly filled.

Chicken meat is extraordinarily versatile for use in cookery. The rich spectrum of possibilities range from light salads and richly flavoured soups to hearty and sophisticated main dishes. Here is a small selection of not exactly everyday, but appetising egg and chicken meat recipes. A rich choice of dishes, from the most basic to the rarest *haute cuisine*, can, of course, be found in the vast array of cookery books available.

Chicken and egg salad.

Chicken and egg salad

250 g roast chicken meat
6 hardboiled eggs
Small tin button mushrooms and/or asparagus pieces
150–200 g mayonnaise. Salt, lemon juice

Cut the chicken meat into strips and the eggs into eighths. Add the well-drained mushrooms and/or asparagus pieces. Season the mayonnaise with salt and lemon juice, pour over and gently fold into the salad. Garnish with egg segments and serve with toast.

Chicken cocktail with pineapple and button mushrooms

150 g roast chicken meat
2 slices of pineapple
1 small tin button mushrooms
Cocktail sauce:
4 tablespoons mayonnaise
4 tablespoons of whipped cream
1–2 tablespoons tomato ketchup
1–2 tablespoons brandy
1 tablespoon sherry
Salt, sugar, lemon juice
To garnish:
lettuce leaves, cocktail cherries

Cut the chicken into strips. Drain the pineapple and mushrooms well and chop finely. Beat together all the cocktail sauce ingredients and season well. Fold in the chicken, pineapple and button mushrooms. Place a small lettuce leaf into a cocktail glass for each person, fill with the cocktail and garnish with a cherry. Serve chilled.

Chicken pie

2 packets puff pastry
750 g skinless chicken meat
1 red and 1 green pepper
1 onion
Bunch of parsley
1 small tin of button mushrooms
Butter, cooking oil
2 eggs
1 cup sour cream, 1/8 l white wine
Salt, pepper, curry powder, paprika powder
For coating:
2–3 tablespoons milk or cream
1 egg yolk

Rinse out a baking dish with cold water and lay a little more than half of the pastry into the dish so that the pastry is 3 cm high at the sides. Cut the chicken into cubes. Briefly brown the chicken with the chopped onion, parsley, green and red pepper and the drained mushrooms in the butter and oil. Whisk together the eggs, cream, salt, pepper, curry powder, paprika powder and white wine.

Distribute the browned meat and vegetable ingredients in the baking dish and then pour over the whisked ingredients. Place a pastry lid on top and brush the edge and the lid with a little water and then press firmly together. Cut a small cross in the middle of the pastry lid with a knife to allow the steam generated during baking to escape. Preheat the oven to 200°C. Brush the pastry with milk or cream and egg yolk. Bake on the lowest shelf of the oven for about 50 minutes.

Spicy wine-pickled eggs

8 hardboiled eggs
1/4 l wine vinegar
1/4 l dry white wine
1/4 l water
1 red onion
1 teaspoon mustard seeds
8 juniper berries
4–5 dried chillies
5 cloves
1 tablespoon salt

1 teaspoon sugar
4 sprigs of dill

Pickled eggs

12 hardboiled eggs
1 l water
50 g salt
3–4 dried chillies
2 sprigs of thyme
1 small onion
2 bay leaves
1 teaspoon caraway seeds

Spicy wine-pickled eggs.

Bring all the ingredients apart from the dill and the eggs to a boil and simmer lightly for five minutes. Put the eggs in an earthenware pot and place the dill sprigs between them. Allow the liquid to cool a little and then pour it over the eggs. Allow to infuse for at least 48 hours. Do not store for more than four or five days, otherwise the eggs will become too acidic. Arrange the eggs in a glass bowl and cover with some of the liquid. Serve with bread, salted butter and mild mustard.

Allow all the ingredients apart from the eggs to boil over a moderate heat. Lightly crack the hardboiled eggs, place them in a glass jar and pour the liquid onto them through a sieve. Allow to cool. Fasten a lid on the jar and leave in a cool place. The eggs taste best when they have been left in the pickling liquid for four or five days. Serve with chunky bread and salted butter. The best way to eat pickled eggs is as follows: carefully halve the shelled egg, loosen and take out the yolk and put it aside. Into the hollow space left by the yolk, place some mild mustard, oil and a few drops of vinegar. Place the yolk upside down on top of this and then place the whole half egg into the mouth in one go. These are also good spiced up with Worcestershire sauce, tomato ketchup or a piquant chutney.

Egg liqueur

15 egg yolks
400 g icing sugar
2 sachets of vanilla sugar
1 pinch of cinnamon
1/2 l cream
1/2 bottle of brandy

Beat the egg yolks and icing sugar until foamy. Bring the cream, vanilla sugar and cinnamon to a boil, then allow to cool and, when lukewarm, gradually whisk into the foam mass.

Egg liqueur.

Stir in the brandy, allow to cool completely and then transfer to bottles.

Egg beer

3/8 l milk
1/8 l cream
1/2 l light beer
100 g sugar
1 knife tip of ground cinnamon
1 knife tip of ground cloves
4 eggs
2 egg yolks
Grated nutmeg

Mix the ingredients together in a saucepan. Place on the hob and, using a moderate heat, beat with a whisk until a thick foamy mass has formed. Quickly pour into four thick-walled glasses (previously swilled out with hot water), dust with a little nutmeg and serve.

Coffee egg punch

4 egg yolks
150 g sugar
1–2 egg whites
1/2 l strong coffee

Mix the egg yolks with the sugar and beat until foamy. Optionally mix in 1–2 egg whites to make the mixture fluffier. Stir the coffee into the egg mixture, heat the whole mixture on the hob and whisk until hot and foaming. Serve in cups.

Appendix

As sources of chickens, chicks and hatching eggs, apart from private and commercial poultry keepers there are also local breed clubs and small-animal breeders.

The feed we use – particularly the ready feed mixes and chick feed – is best bought from agricultural suppliers or from a local mill. From mills, we also buy inexpensive waste cereal, which we can give to the birds as a feed supplement, particularly in winter, mixed into the henhouse litter.

We source our feed and water containers from agricultural suppliers, who also offer specialist advice and suggestions.

Internet searches are also helpful, of course.

In the event of infectious diseases in the flock, it is best to contact a vet or the local veterinary authorities.

Index

www.ingramcontent.com/pod-product-compliance
Ingram Content Group UK Ltd.
Pitfield, Milton Keynes, MK11 3LW, UK
UKHW050137280726
14058UKWH00006B/691

9 781789 184129